The Naval Aviation Guide

FOURTH EDITION

The
Naval Aviation
Guide

Edited by Captain Richard C. Knott, USN

Naval Institute Press
Annapolis, Maryland

Library of Congress Cataloging-in-Publication Data
Main entry under title:

The Naval aviation guide.

 Rev. ed. of: The naval aviation guide / Malcolm
W. Cagle. 3rd ed. c1976.
 Includes index.
 1. United States. Navy—Aviation. 2. United
States. Navy—Officers' handbooks. I. Knott,
Richard C. II. Cagle, Malcolm W. Naval aviation guide.
VG93.N34 1985   358.4'0023'73   85–15535
ISBN 0–87021–409–8

Printed in the United States of America

Contents

Preface

This edition of the *Naval Aviation Guide* is designed to be an informative, concise, and useful tool of the naval aviation profession. It brings together a large amount of information which is, for the most part, available in a variety of other sources but which is more accessible under one cover. Every major facet of naval aviation is treated here, from its origins and development to its organizational relationships and working parts. A thirteen-part appendix provides additional ready-reference material to help answer day-to-day questions or simply to satisfy the reader's curiosity in a specific area.

The new *Naval Aviation Guide* is an introduction for those considering that first step into the exciting world of naval aviation as well as for those already earning their wings of gold. The guide should prove especially helpful to pilots and naval flight officers beginning their first operational tours, and it should serve as a handy reference source for old hands. And for anyone, military or civilian, professional aviator or armchair pilot, who wants to learn more about the profession, this guide is designed to present a clear picture of an important branch of naval warfare.

Most of the text and the illustrations in this book are completely new. A few chapters are revisions of material in previous editions of the guide. Information in each revised chapter has been carefully edited and updated to reflect recent developments. In every instance, authors and revisers were selected for their expertise in the subject area.

It is with great appreciation that I recognize those who labored on this project, who spent considerable time and effort gathering the most relevant information and presenting it in the most readable form. In addition to the individual contributors to this book, listed at the end, I am also indebted to Mr. Robert Lawson, editor of *The Hook*, for granting permission to reprint the information on aces and MiG killers that originally appeared in his magazine.

The Naval Aviation Guide

1

Naval Aviation: The Legacy

By Captain Albert L. Raithel, USN (Ret.)

THE PIONEERS

At its inception, naval aviation, the projection of naval warfare into the air, was a revolutionary idea with far-reaching effect. With all the advantages of hindsight, we today can appreciate the impact naval aviation has had on our national defense and the wisdom of those who spurred its development. But this was not always the prevailing view. Naval officers who championed the idea of naval aviation did so at the risk of their careers and their credibility. Yet they persisted, and it was only through their foresight and determination that naval aviation eventually came into its own.

It is sometimes difficult to say precisely when an idea is born, and so it is with naval aviation. Officially, its birthday is 8 May 1911, for it was on that day that Captain Washington Irving Chambers prepared the requisition to purchase two Curtiss biplanes, the navy's first aircraft.

Orville and Wilbur Wright had already flown an aeroplane in

The Legacy (Artist: R. G. Smith)

the sand dunes of North Carolina in 1903. Five years later a joint board convened at Fort Myers, Virginia, to evaluate the Wright aircraft for possible military use. Lieutenant George C. Sweet was the navy member of the board, and Naval Constructor William McEntee was detailed as an observer. The demonstrations ended in tragedy when the machine crashed and Lieutenant Thomas Selfridge of the U.S. Army died. Despite this sobering turn of events, Lieutenant Sweet believed the aeroplane had potential and recommended the purchase of four aircraft for adaptation to naval use. However, the secretary of the navy, who had also observed the trials, was more impressed by the crash, and no further action was taken.

In 1910 the U.S. Navy took its first small step in the direction of naval aviation when it appointed Captain Washington Irving Chambers, a nonflier, to keep himself informed on the progress of aviation. In the fall of that year, he met the inventor-aviator Glenn Curtiss and Eugene Ely. Curtiss and Chambers soon embarked on a dramatic series of demonstrations to show the adaptability and usefulness of the aeroplane to the navy.

Could planes actually be launched from a ship and recovered at sea? The first question was partially answered on 14 November 1910, when Ely, a civilian associate of Curtiss, successfully flew from a platform hastily erected on the cruiser *Birmingham* at Norfolk, Virginia. This flight generated enormous interest, both in the United States and abroad, but no navy funds for the purchase of aircraft were forthcoming.

At about this time, Curtiss was in the process of moving his aviation camp to California for the winter, and he and Captain Chambers again teamed to promote naval aviation. Curtiss, knowing that funds were not available, offered to instruct a naval officer to fly a Curtiss plane, at no cost to the government. Lieutenant T. G. "Spuds" Ellyson, a submariner, was ordered to the Curtiss camp at North Island to become the first U.S. naval aviator.

In January 1911 a special platform was installed on the armored cruiser *Pennsylvania*, and on the eighteenth Ely landed on it. The plane was launched and Ely returned to the beach the same morning. Meanwhile Curtiss, with the assistance of Ellyson, had developed the world's first practical seaplane. A month later he landed it alongside the *Pennsylvania* and was hoisted aboard by the ship's boat crane.

Following the success of these demonstrations, funds were finally made available for aviation, and the navy purchased two aircraft from Curtiss. The first was called the Triad, because it could operate from the land, from the sea, and in the air. Later an airplane was purchased from the Wrights. These two manufacturers each agreed to train a pilot and a mechanic to fly the aircraft they sold to the navy. Lieutenants John Rodgers and John H. Towers thus became the navy's second and third aviators, respectively.

Captain Chambers next made arrangements with the superintendent of the Naval Academy for the first naval air station, which was established at Greenbury Point, near Annapolis, Maryland, on 27 June 1911. A wooden hangar was erected, trees were felled, and a swamp was partially filled to provide a landing field. Much valuable experience was gained during the stay at Greenbury Point. The first night operations were conducted there. In the winter of 1911–12 the fliers moved back to North Island, where the climate was more suitable for flight operations. With the return of spring, Annapolis again became the scene of activity, and in May 1912 First Lieutenant A. A. Cunningham became the first marine assigned to aviation duty.

Thus far, however, the aeroplane had not proven itself to be an asset in the conduct of naval warfare. Consequently, in January 1913 Captain Chambers arranged to move the aviation camp to Guantanamo Bay, Cuba, for the first practical tests with the fleet. Five planes and all the navy's aviators and student aviators deployed. Their objectives were to prove the value of aircraft for scouting, to further test the capability of aircraft to detect submerged submarines and mines, and to indoctrinate as many fleet officers as possible in the potentialities of aviation in support of the fleet.

A decision was now made to locate the aeronautic station at the recently closed navy yard at Pensacola, Florida. Lieutenant Towers took over as officer in charge of the aviation camp, and

Lieutenant Commander H. C. Mustin became commandant of the aeronautical station. The training of new aviators soon commenced.

Shortly after arriving at Pensacola, the majority of aviation personnel and equipment was deployed to Mexico under Mustin's overall command. Towers and his detachment were sent to Tampico aboard the *Birmingham*, while Lieutenant Patrick N. L. Bellinger and his detachment went to Vera Cruz on board the *Mississippi*, where they flew scouting missions for the troops ashore. On one mission Bellinger drew rifle fire, and his plane sustained the navy's first aircraft combat damage.

Between 1914 and 1917 the navy concentrated on the problem of taking planes to sea on existing warships. The *North Carolina* was fitted with a catapult, and on 5 November 1915 Lieutenant Commander Mustin was successfully launched with the ship under way in Pensacola Bay. In addition to the *North Carolina*, the *Huntington* and *Seattle* were also fitted with catapults and planes.

Following his return from the Mexican expedition, Towers was sent to the Curtiss facility at Hammondsport, New York, to assist in testing the flying boat *America*, which had been designed and built to attempt the world's first flight across the Atlantic. Tests showed that it was a qualified success, but the outbreak of war in Europe led to its cancellation. *America*, however, served as the prototype for the flying boats that dominated maritime patrol activities in the American and British navies during World War I and immediately after the war.

THE GREAT WAR

In August 1916 several events occurred that laid the foundation for the rapid expansion of naval aviation when the United States finally entered the war. The Appropriations Act of 29 August 1916 provided the sum of 3.5 million dollars for aeronautics and authorized the formation of a naval flying corps and a naval reserve counterpart. Soon the navy had its first large-scale production contract for thirty Curtiss N-9 training seaplanes, and the personnel necessary to man a wartime force had to be recruited. The act also provided for a commission to make recommendations for expansion of naval aviation shore facilities to support wartime requirements.

With the approval of the secretary of the navy, a private organization known as the National Aerial Coast Patrol Commission was formed. Among other things, this group organized aerial coast patrol units at Yale, Princeton, Harvard, and Columbia universities. Personnel of these units supplemented the small cadre of regular officers and provided the bulk of the leadership when naval aviation commenced active wartime flying.

On 6 April 1917, the day war was declared, the strength of naval aviation, navy and marine corps combined, was 48 officers, 239 enlisted men, 54 airplanes, 3 balloons, and 1 blimp. Only one station at Pensacola, Florida, was in commission, and it is an understatement to say that naval aviation was ill prepared for its wartime role. But changes were quickly made to remedy this situation. Hurried consultation with our British, French, and Italian allies formed the basis for initial planning.

German submarine warfare posed the greatest immediate threat to the success of Allied arms. Top priority was given to programs to contain the U-boats and permit supplies to flow to the United Kingdom and Europe. For naval aviation this meant the production and deployment of antisubmarine aircraft. Curtiss had pioneered long-range flying boat design with the *America* and had contracts with the Royal Navy for flying boats of the Large America (H-12) model. The chief problem to be solved was the provision of a high-powered, reliable, standard-design engine, capable of being mass produced.

In late May and early June 1917, J. G. Vincent of the Packard Company and E. J. Hall of Hall-Scott collaborated to design the famous Liberty engine for mass production. It was so engineered that it could be produced in four-, six-, eight-, or twelve-cylinder versions, depending on the power requirements of a particular aircraft. The Liberty became the prototype for design of all the navy's wartime patrol and bombing aircraft models, the H-16, HS-1 and -2, F-5-L, DH-4, and the plane used by the Northern Bombing Group Capronis. The Liberty was a sound engine, fortunately for naval aviation, for there was no alternative engine available.

A decision was also made to establish an aircraft factory to be owned by the navy, which would assure part of the navy's aircraft production, serve as a source of production cost data, and give the navy an in-house capability for producing experimental designs.

A contract was let on 6 August 1917 for the naval aircraft factory located at the Philadelphia Navy Yard. The plant was completed 110 days later, and the first and second production aircraft were shipped abroad on 2 April 1918.

In May 1917, primary lighter-than-air training was shifted from Pensacola to a facility owned by the Goodyear Company in Akron, Ohio, while both advanced lighter- and heavier-than-air training continued at Pensacola.

A preflight school was established at the Massachusetts Institute of Technology. Soon all officer candidates, both flight and ground trainees, were receiving their initial training there. Later, as officer programs expanded, additional preflight schools were established at the University of Washington and at the Dunwoody Institute in Minneapolis, Minnesota.

Enlisted personnel required to support the rapid expansion of naval aviation were initially trained at the operating air stations. Aviation schools were later established at the Great Lakes Naval Training Center, while technical specialty training was conducted at various private industrial facilities and the navy gas-engine school at Columbia University.

The first organized unit of the United States armed forces to deploy overseas following the declaration of war was the navy's First Aeronautical Detachment, under command of Lieutenant Kenneth Whiting. Arriving in France on 5 June 1917, personnel of the unit received pilot and aircrew training at French facilities. They provided the nucleus of trained personnel for the first four American naval air stations established in France. Flight-training facilities were also provided by the Italian government at Lake Bolsena, which was made a naval air station in February 1918.

Patrol operations were inaugurated on both sides of the Atlantic shortly after the declaration of war. An antisubmarine patrol station was established at Punta Delgado, Azores, by the First Marine Aeronautic Company, and the naval air station at Coco Solo in the Canal Zone was commissioned to contribute to the defense of the Panama Canal. Plans for strategic bombing operations against German industrial and naval targets were put into effect. The Northern Bombing Group, to be composed of night- and day-bombing squadrons, commenced operations several months before the war was over. British squadrons that included some American pilots had concentrated their early efforts against German submarine bases in Belgium. Later operations, particularly those of the Marine Day Wing, supported the Allied land offensive operations in October and November 1918.

Two additional strategic bombing programs failed to develop fully. One involved towing seaplane bombers on lighters within range of German bases. Some lighter stunts, as they were referred to, were carried out successfully by the British. Another innovative program involved the use of self-propelled sea sleds, each of which would carry and launch one aircraft to bomb German targets. The sea sled's shallow draft permitted it to operate over and inside the German defensive minefields. Successful tests were conducted, but the end of the war arrived before sleds could be put into operation.

Other plans for 1919, never carried out, called for the construction of six aircraft carriers and the formation of a southern bombing group to operate from Italian bases against the Austrian navy.

World War I did much to establish an early role for naval aviation in bombing and antisubmarine warfare. Navy and marine corps aircraft logged over two million nautical miles on patrols from U.S. continental stations and almost eight hundred thousand nautical miles on overseas patrols and bombing missions. More than one hundred thousand pounds of bombs were dropped on German bases and naval targets. Twenty-five attacks were made on German submarines, twelve of which were reported damaged.

A new confidence emerged from wartime feats of navy and marine corps aviators. Ensign Charles H. Hammann was awarded the Medal of Honor for his spectacular rescue of a fellow aviator from the Adriatic Sea in the face of strong enemy opposition. Second Lieutenant Ralph Talbot and Gunnery Sergeant Robert G. Robinson were each awarded the Medal of Honor for extraordinary heroism in their running fight with twelve enemy aircraft.

Lieutenant (j.g.) David S. Ingalls became the navy's first "ace," destroying four enemy aircraft and one or more enemy balloons.

In short, the expansion of naval aviation was extraordinary during the nineteen months of our involvement in World War I, as table 1-1 indicates.

Table 1–1. Navy and Marine Corps Aviation, World War I

	6 April 1917	11 November 1918
Personnel		
Officers	48	6,998
Enlisted men	239	32,873
Overseas	—	approximately 18,000
Air Stations		
U.S. continental	1	12
Overseas	—	31
Aircraft Inventory		
Seaplanes/flying boats	51	1,865
Landplanes	3	242
Dirigibles	1	15

The armistice ending World War I came on 11 November 1918 accompanied by public demand for cessation of huge wartime expenditures. Demobilization proceeded at a rapid pace with drastic reductions in personnel and funds. Despite this, the navy continued to work on the problem of integrating aviation into the fleet, a goal that had been put aside during the massive wartime antisubmarine and bombing effort.

In January 1919 a squadron of flying boats conducted exercises with the fleet off Guantanamo Bay, Cuba. These aircraft joined a detachment of Sopwith Camels, flying from a platform over the number two turret of the battleship *Texas*, and six Kite Balloons and their crews, based ashore. The flying boats, supported by the minelayer *Shawmut*, which was operating as a seaplane tender, demonstrated that they could travel with the fleet and provide long-range services as required. The success of the Sopwith Camel detachment led to the formation of the Atlantic Fleet Ship Plane Division, which operated aircraft from turret platforms on four battleships.

FIRST FLIGHT ACROSS THE ATLANTIC

During World War I Admiral David W. Taylor had initiated work on long-range flying boats capable of crossing the Atlantic nonstop and possessing excellent seakeeping capability. The Navy Curtiss (NC) planes were designed by three naval constructors, H. C. Richardson, J. C. Hunsaker, and G. C. Westervelt, in collaboration with Glenn Curtiss. With a wingspan of 126 feet, a 45-foot hull length, and three Liberty engines, the big plane was ready for its first flight in October 1918.

The war ended before the NCs were ready for service, but development continued and the planes gained another engine. During tests the four-engined NC-1 carried fifty-one persons aloft, setting a world record.

A decision was made to use these aircraft to attempt the world's first Atlantic crossing by air. On 8 May 1919 three NCs, led by Commander John Towers, departed the naval air station at Rockaway Beach, New York, bound for Plymouth, England, via Halifax, Nova Scotia, Trepassy Bay, Newfoundland, the Azores, and Lisbon, Portugal. All three aircraft left Trepassy Bay as planned, but navigation difficulties forced two of them to land at sea. Towers sailed NC-3 into the Azores, but extensive damage prevented further flight. The NC-1 turned over and sank when taken in tow.

NC-4, commanded by Lieutenant Commander A. C. Read and flown by Lieutenant (j.g.) Walter Hinton and coast guard Lieutenant Elmer Stone, made it safely to Lisbon, Portugal, on 27 May, marking an important milestone in aviation. She then completed the flight to Plymouth, eliciting much public acclaim from both sides of the ocean.

YEARS OF GROWTH

In the spring of 1919 the General Board of the navy recommended the establishment of a naval air service and the fullest development of fleet aviation. The General Board also recommended that the

The Navy Curtiss NC-4 was the first aircraft to fly across the Atlantic. Taking off from Trepassy Bay, Newfoundland, and stopping in the Azores, it landed in Lisbon, Portugal, in May 1919. (U.S. Navy)

collier *Jupiter* be converted to an aircraft carrier for temporary use and experiments. The USS *Langley* (CV 1) was commissioned on 20 March 1922. Later that year Lieutenant Commander Godfrey deC. Chevalier made the first arrested landing, and American carrier aviation was born.

Lighter-than-air proponents meanwhile had been advocating a rigid airship program for several years. The wartime success of Allied nonrigid airships and the success of the German zeppelins in scouting supported their arguments. The availability of inert helium gas as a substitute for the highly inflammable hydrogen greatly reduced the risk of such operations. Encouraged by Admiral Sims, wartime commander of U.S. Naval Forces in Europe, Congress agreed, and the appropriations for fiscal year 1920 contained funds for building one rigid airship in this country and buying one abroad.

The parts for the first of these, the *Shenandoah* (ZR-1), were fabricated at the naval aircraft factory, and the airship was assembled at the new naval air station at Lakehurst, New Jersey. The second ship, the partially completed R-38, was to be purchased from the British. Trials of the R-38 began in June 1921; on her fourth flight, prior to final acceptance, she broke in two and fell in flames into the Humber River, near Hull, England.

Under the Versailles Treaty, the United States was entitled to two German zeppelins as a reparation. These and other ships scheduled for delivery to the Allies were destroyed by their crews. Subsequently, the Germans built the United States one new zeppelin. This became the small but highly successful *Los Angeles* (ZR-3).

While progress was being made in the carrier and lighter-than-air programs, long-range patrol aviation continued its slow, steady record of achievement. Metal hulls and more reliable engines were incorporated in models that were, to a large extent, improvements to the basic F-5-L design.

A significant organizational change, which had a lasting effect on the development of naval aviation, was the creation of the Bureau of Aeronautics. Prior to 10 August 1921 the various command and material requirements of naval aviation had been accommodated by extensive and at times exhaustive coordination between the Office of the Director of Naval Aviation and the material bureaus of the navy. The Bureau of Construction and Repair provided the airframes, the Bureau of Engineering engines and radios, the Bureau of Navigation instruments and navigation equipment, and the Bureau of Ordnance the weapons.

The new organization brought the majority of aeronautical material under the cognizance of the Bureau of Aeronautics. Spurred by the energetic leadership of its first chief, Rear Admiral William Moffett, soon to qualify as the navy's first naval air observer, and his deputy, Captain Henry C. Mustin, the navy's senior aviator, rapid progress was made integrating aviation into the navy.

In 1925 Commander John Rodgers and a crew of four in a PN-9 flying boat attempted a nonstop flight from San Francisco to Hawaii. After flying more than three quarters of the way across, they were forced down by lack of fuel. Fashioning a sail from wing fabric, and using the flooring for leeboards, they sailed the aircraft tail first for the remaining 450 nautical miles to the island of Kauai in ten days. The object of an extensive search, they had almost been given up for lost. The flight of 1,841 miles was accepted as an international distance record for class C seaplanes that remained unbeaten for almost five years.

The USS *Saratoga* (CV 3) was converted into an aircraft carrier from a partially built battle cruiser. (U.S. Navy)

The Washington disarmament treaties of 1922 had an important impact on naval aviation. U.S. battle cruisers then under construction were in excess of treaty limitations and had to be disposed of. At the same time, the treaties allowed the United States two aircraft carriers of not more than 36,000 tons each. Two of the battle cruisers were converted into the carriers *Lexington* (CV 2) and *Saratoga* (CV 3), both commissioned in late 1927.

A highly successful aerial survey of Alaska was made in 1926 by navy amphibious aircraft supported by the tender *Gannet*. Also during the twenties, Commander (later Rear Admiral) Richard E. Byrd gained international fame and in the process developed many of the cold-weather techniques later used during high-latitude operations in wartime. His accomplishments included the aerial exploration of northern Greenland, the first flight over the geographic North Pole with naval aviation pilot Floyd Bennett, and a major expedition to Antarctica in 1929, where he made the first flight over the geographic South Pole. This latter expedition became the forerunner of a series of U.S. expeditions to Antarctica supported by naval aviation to this day.

The decade of the twenties stands out in the history of naval aviation as a period of phenomenal growth. Although it was generally characterized by declining appropriations, the navy devoted a steadily growing share to aviation. By 1929 three carriers were in commission, patrol squadrons supported by seaplane tenders were performing scouting operations for the fleet, aircraft were based aboard battleships and cruisers for scouting and gunfire spotting, and the marine corps had learned many valuable lessons about providing air support for expeditionary forces.

Aircraft flew higher, faster, farther, carrying heavier and heavier loads. Many world aviation records were made by naval aviators. Accurate dive-bombing became an accepted tactic that was to prove decisive in future naval battles. Torpedo attack doctrine, patrol and scouting techniques, and advanced-base operations were also developed effectively.

The Great Depression that began in 1929 took its toll on naval aviation. Operating tempos and support funds were significantly reduced. Air stations at the low end of the funding priority list fell far behind in their capability to support authorized programs.

The *Ranger* (CV 4), the first U.S. Navy ship to be designed and built as a carrier, was authorized and her keel laid in 1931, but subsequent appropriations were insufficient to procure planes for her air group. Patrol aircraft assigned to the Canal Zone, Hawaii, and the Asiatic Fleet were reduced to provide funds for the *Ranger's* planes.

Two reserve air stations were commissioned in the early 1930s at St. Louis, Missouri, and Opalocka, Florida. These were largely financed by local rather than federal funds. Only construction of the major fleet lighter-than-air base at Sunnyvale, California, made real progress during this period. It was built on a thousand acres sold to the federal government by local interests for one dollar.

Lighter-than-air activities continued to attract wide public attention. The *Los Angeles* (ZR 3) had established a sound record of performance, and valuable operating experience had been gained during her early service.

The *Akron* (ZRS 4), the first of the giant dirigibles, was commissioned on Navy Day 1931. To overcome a recognized defensive weakness, she was equipped with a hanger and the ability to carry as many as four fighters. These were equipped with overhead hooks and were recovered aboard through use of a trapeze.

By March 1933 the *Akron* had made seventy-three flights for almost seventeen hundred hours. On the night of 3 April, while engaged in radio-direction-finder calibration, she was caught in a violent storm off the coast of New Jersey and crashed. Only three of her crew of seventy-six survived. Among those who died was Admiral William Moffett, the first chief of the Bureau of Aeronautics.

The *Macon* (ZRS 5) was commissioned in June 1933 and assigned to the new base at Sunnyvale, California. But she too was lost in a storm in February 1935, and was not replaced. The day of the giant dirigible in the navy had passed, though nonrigid airships would play an important part in the antisubmarine campaigns of World War II.

As more aircraft carriers joined the fleet and exercises more nearly approached conditions that would be encountered in war, it became apparent that carrier planes with increased performance capabilities would have to be developed. Some of the great aircraft of World War II began to evolve during this period.

The need for more capable tenders for patrol aviation also be-

The airship USS *Macon* (ZR 5) over New York City in the early 1930s. (U.S. Navy)

came apparent, and the idea of building small carriers for use in support roles was first considered.

Personnel shortages continued to plague the fleet. The Naval Academy was unable to supply sufficient officer graduates as student naval aviators, and training of enlisted pilots had been stopped in 1934 as an economy measure. In fact, the enlisted naval aviation pilot (NAP) program had not been continuous since it was first instituted. After the first two classes of enlisted pilots graduated from Pensacola in 1917, the formal training program was suspended.

The training of enlisted pilots had resumed on a somewhat regular basis in 1920, and the designation naval aviation pilot (NAP) came into being. But the training of NAPs proceeded by fits and spurts, ending in 1933 and being reinstituted again in 1936. The Aviation Cadet Act of 1935 once more established a naval reserve program that would provide the experienced cadre on which naval aviation's wartime expansion would draw.

Fleet exercises during this period provided extensive experience in planning and execution of fast carrier strikes against shore installations and in the development of circular screens for antiaircraft defense of carrier formations. The vital role of aviation in support of landing operations by the Fleet Marine Force was recognized, and a high degree of coordination was developed. A healthy spirit of competition among naval aviation units further enhanced combat skills.

WAR CLOUDS

Days after the start of the war in Europe, the Atlantic Squadron was directed to establish "neutrality patrols" over areas as far east as the sixty-fifth meridian and as far south as Trinidad. The objective of the "neutrality patrols" was to make the task of the German U-boats as difficult as possible.

Hardware became a matter of immediate concern. In addition to the air groups training for the *Ranger* and the new carrier *Wasp* and a few planes in the Canal Zone, the Atlantic Squadron had only twenty-five scout planes based on battleships and cruisers and fifty-four patrol planes. The rest of naval combat aviation was in the Pacific. On 16 May 1940 President Roosevelt called for an army and navy aircraft expansion program of fifty thousand planes, and a year later he declared that an unlimited state of national emergency existed.

In July 1941 the United States took over the garrisoning of Iceland from the British, and shortly thereafter Reykjavík became a base for routine patrol flights over the Atlantic convoy routes. In October the navy received its first land-based patrol planes, and on 1 November the coast guard was transferred to the operational control of the navy.

As the war approached, naval aviation could muster 6,750 navy, marine corps, and coast guard pilots; 1,874 ground officers; 21,678 enlisted men; 5,260 aircraft of all types, including trainers; 7 large and 1 small aircraft carriers; 5 patrol wings; 2 marine air wings; and a few advanced air bases.

IN THE THICK OF IT

On 7 December 1941 Japan attacked Pearl Harbor to destroy or cripple the Pacific Fleet. She calculated that the move would free her to extend control over the southern resources area, including the Netherlands East Indies, the Philippines, and Southeast Asia. But the Japanese attack found all the American carriers at sea. This factor alone gave Americans cause for hope in the desperate months ahead.

In reality, the events of 7 December hastened a transition that had been slowly taking place in naval thinking. The battleships of the Pacific Fleet were sunk or damaged, leaving the carriers as the major capitol ships of the fleet. The Japanese attack was followed by a declaration of war on the United States by Germany and Italy on 10 December.

The situation in the Pacific in early 1942 was grim. American forces in the Philippines were overwhelmed by a sustained Japanese assault. The naval forces that survived the initial attacks regrouped and fought a series of delaying actions as they fell back through the Netherlands East Indies to Australia, eastern New Guinea, and the New Hebrides area. The PBY Catalinas of Patrol Wing 10 performed courageously in the face of great odds to provide scouting and patrol reports to Allied commanders and to attack the enemy whenever possible. But it was to no avail. The *Langley*, converted to a seaplane tender, was sunk while ferrying a load of fighters to defenders in the Netherlands East Indies. Seemingly, nothing could stop the Japanese advance.

The first of a series of actions planned to keep the enemy off balance took place in February 1942. In the first U.S. offensive carrier operation, task forces built around the *Enterprise* and *Yorktown* conducted raids against the Japanese-held Gilbert and Marshall islands. Compared with later operations, this was a small one, but it gave the navy and the nation a morale boost when it was sorely needed.

In March the *Enterprise* launched attacks on Marcus Island within one thousand miles of Japan proper, and in April sixteen B-25 bombers of the Army Air Force, led by Lieutenant Colonel

James H. (Jimmy) Doolittle, were launched from the *Hornet* and attacked Tokyo and three other Japanese cities.

CORAL SEA AND MIDWAY

Following the initial success of their original plan of expansion into the southern resources area, and partially spurred to action by the Doolittle raid, the Japanese commenced another series of operations to secure a defensible outer perimeter. The first of these moves was designed to capture Port Moresby in southeastern New Guinea.

Now came the first real test of U.S. carrier aviation. It was an especially significant event in the history of naval warfare because the outcome was decided solely by air power. Opposing surface forces were never within visual or firing range of one another.

In a series of engagements that later became known as the Battle of the Coral Sea, the Japanese lost the light carrier *Shoho*, while another carrier of theirs, the *Shokaku*, suffered damage at the hands of American pilots. For their part, the Japanese inflicted damage on the U.S. carriers *Lexington* and *Yorktown*, and the former had to be abandoned and sunk. Nevertheless, enemy plans to invade Port Moresby had been foiled, and the Japanese retired to lick their wounds.

Less than a month later, almost five thousand miles to the north of the Coral Sea, the Japanese planned another major thrust. Their objective was to occupy Midway Island in the Central Pacific and several of the Aleutian Islands to establish an outer defense line to the east and northeast of Japan. From these bases, air power could be projected some thirteen hundred miles, a radius that included the Hawaiian Islands.

The Japanese knew that all available U.S. carrier strength had been engaged at Coral Sea. They were aware of the relatively strong defensive effort the United States was mounting in the South Pacific to secure the sea lines of communication to Australia, and they did not anticipate strong opposition to their new moves in the central Pacific. Accordingly, they set 6 June as the date for the occupation of Midway.

Cryptanalysis permitted the United States to intercept and read

Japanese message traffic, and armed with the knowledge thus gained, the Pacific Fleet commander, Admiral Chester W. Nimitz, ordered all available forces to the Midway area.

By the fourth of June, the opening day of the battle, the two tiny islands at Midway were covered with 120 army, navy, and marine corps aircraft, weapons of all sorts, and over thirty-five hundred defenders from the three services. At sea, Task Force 16, with the *Enterprise* and *Hornet*, and Task Force 17, with the *Yorktown*, rendezvoused northeast of Midway.

In opposition, the Japanese had put together one of the most powerful naval forces ever assembled. Led by a four-carrier strike force and supported by heavy units of the Main Body, the Occupation Force steamed towards Midway with confidence.

A PBY Catalina from the island located ships of the Occupation

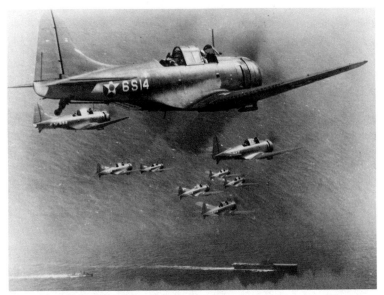

The Battle of Midway was the turning point of the war in the Pacific. Here Douglas SBD Dauntless dive-bombers, which helped rout the enemy in this critical encounter, fly over the USS *Enterprise* (CV 6), one of the three U.S. carriers that participated in the battle. (U.S. Navy)

Force six hundred miles to the west on June 3, and on the following day another of these patrol aircraft sighted enemy planes from the strike force inbound to Midway. Land-based aircraft engaged in defensive actions and made torpedo and bombing attacks on the Japanese Strike Force. Diverted by these attacks, the Japanese carriers were not prepared to meet the full weight of American carrier aircraft that followed. Beginning at 0930 with the attack by Torpedo Squadron 8, and continuing with torpedo and dive-bombing attacks, the morning raids left three Japanese carriers, the *Akagi*, *Kaga*, and *Soryu* heavily damaged. They would all sink within the day, the *Soryu* being finished off by torpedoes from the American submarine *Nautilus*.

In two Japanese counterattacks commencing about noon, the *Yorktown* was severely damaged and abandoned. Later American planes found the fourth carrier of the Japanese strike force, the *Hiryu*, and sank it in a dive-bombing attack. As a result of the loss of all four carriers of the strike force, the Japanese commander, Admiral Isoroku Yamamoto, ordered the abandonment of the operation.

Further attacks by Midway-based planes on the fifth and by *Hornet* and *Enterprise* dive-bombers on the sixth sank one cruiser and damaged another cruiser and a destroyer.

The *Yorktown* was reboarded by a salvage party on the fifth, and damage-control measures were progressing satisfactorily when a brace of torpedoes fired by a Japanese submarine mortally wounded her.

As a result of the Battle of Midway, Japanese eastward expansion was halted. For them the battle was a disaster. The loss of four of their finest carriers, 250 planes, and 100 of their most experienced pilots deprived them of the striking power they needed to maintain their momentum of conquest. From the date of this battle, the balance of power in the Pacific shifted steadily to the United States and her allies.

To the north, the Japanese had little better luck. Following an inconclusive attack on Dutch Harbor, Alaska, their forces occupied Kiska and Attu islands in the western Aleutians, beginning on 6 June. Due to the vagaries of the Aleutian weather, Japanese occupation of these islands was not detected until 10 June. It was the lot of Patrol Wing 4 PBY Catalinas to carry the war to the enemy in the Aleutians.

Beginning on 11 June and operating from Dutch Harbor and the tender *Gillis* at Nazan Bay, Atka Island, the Catalinas conducted an around-the-clock operation that came to be called the Kiska Blitz. PBYs carried out continuous bombing and strafing attacks on ships and installations in Kiska harbor, refueling, rearming, and returning to the battle as rapidly as the limited facilities allowed. At times the lumbering PBYs were used as dive-bombers at speeds up to 250 knots. In four days the limited supply of bombs, torpedoes, ammunition, and gasoline was exhausted.

The Japanese remained, but they had been badly shaken. Aircraft of Patrol Wing 4 and the army's Eleventh Air Force continued to attack and harass the Japanese until they were finally forced to abandon the Aleutians in the summer of 1943.

GEARING UP FOR VICTORY

At home, training shifted into high gear. New training stations and facilities that permitted increased operational flying at coastal air stations were opened in the Mid West. In March 1942 the aviation pilot rating for enlisted personnel was reestablished, and it was not eliminated from the rating structure until 1948. A number of NAPs were commissioned during the war.

As the war progressed, two side-wheel excursion steamers were converted into the training carriers *Wolverine* and *Sable*. They were based in the Great Lakes, where they operated without interference from enemy submarines. A program was also inaugurated to train nonpilot navigators. Graduates were assigned to multiengine flying activities throughout the fleet. The first women to qualify for navy wings were six navigation instructors who became naval aviation observers (navigation) in the women's reserve. Other programs trained naval aviation observers for radar and tactical operations. Each program had its own distinctive insignia, all of which were abolished in 1947 when these personnel were authorized to wear the insignia of naval aviation observers.

Aviation ground officers performed a wide variety of duties.

Starting with direct commissioning of aeronautical engineers in 1941, the program quickly expanded along with other segments of naval aviation. Initially training was conducted at the naval air station in Anacostia, D.C., but the school was soon moved to the station at Quonset Point, Rhode Island, where the air combat information school was also established.

As the increasing demand for nonflying personnel taxed available resources to the limit, women, known as WAVES (women accepted for voluntary emergency service), began to enter the navy in significant numbers. Many filled clerical and professional billets in various divisions of the Bureau of Aeronautics, while others were trained in aeronautical specialties.

Shipbuilding support for wartime expansion began in 1941, when the *Essex* (CV 9) was being constructed. She was the first of the twenty-four large carriers that were the backbone of our fast carrier task forces from 1943 on. Also in 1941, the experimental conversion of a merchant hull to a small aircraft carrier resulted in our first escort carrier, the *Long Island*. During the war, over one hundred escort carriers were constructed for the U.S. Navy and the Royal Navy. A third class of carrier, the light carrier, converted from light cruiser hulls in 1942–43, increased the number of fast carriers. There were nine ships in this class.

Fifteen large and sixty-two small seaplane tenders served in the fleet during the war, as did sixteen types of aviation logistic-support ships.

As mobilization at home intensified, a huge American invasion fleet crossed the submarine-infested waters of the Atlantic and on 8 November 1942 landed troops at three locations on the coast of French Morocco. Hostilities ceased three days later, and navy patrol squadrons immediately began operations from Casablanca and Port Lyautey.

DEALING WITH THE SUBMARINE MENACE

As 1942 came to a close, the antisubmarine picture had changed considerably. From sheer desperation and improvisation in the early months of the war, effective air and surface countermeasures had contained the submarine menace from Iceland to Newfoundland, down the Atlantic coast, through the Caribbean and West Indies, and from the Canal Zone along the coast of South America to Brazil. The approaches to Gibraltar, the west coast of Africa, and the Ascension Island area were also covered by air patrols.

New high-speed, long-range, land-based patrol planes, the PV-1 Ventura and PB4Y-1 Liberator, were added to the fleet. They extended search areas and forced the Germans to move their submarine operations to midocean.

New weapons and equipment were introduced as antisubmarine forces grew stronger. Sonobuoys allowed the tracking of submerged submarines, and magnetic anomaly detectors (MADs) provided the localization capability needed for attacks on submerged submarines. Retro-rockets allowed attacks on the on-top mark of a MAD contact, as the rearward thrust of the rockets' path compensated for the forward speed of the aircraft.

Probably the most important innovations were the X-band radar and the Leigh light. Submarines used the night hours to run on the surface. They could thus use their relatively high surface speed to position for attack and at the same time keep their batteries charged for maximum submerged performance if evasion were necessary. With the advent of small, powerful radars and searchlights, the effectiveness of nighttime air antisubmarine operations was significantly increased. Once the Allies had gained the upper hand in the Battle of the Atlantic, it was never relinquished.

ACROSS THE PACIFIC TOWARD JAPAN

After Midway the Pacific War, from the U.S. standpoint, entered the offensive-defensive phase. At the beginning of the Solomons campaign, an uncompleted Japanese airfield at Lunga Point, Guadalcanal, was captured and quickly placed into operation as Henderson Field. Possession of this airfield, at times hotly contested, would be the key factor in the ultimate Allied success in the battles that lay ahead.

On 24–25 August two task forces with the *Saratoga* and *Enterprise* engaged a Japanese force attempting to reinforce Guadalcanal. In the ensuing battle, the Japanese carrier *Ryujo* was sunk by planes from the *Saratoga*. The Americans sustained no losses in this battle, although an attack on the *Enterprise* resulted in heavy damage.

On the morning of 14 September, the Seventh Marine Regiment departed from Espíritu Santo for Guadalcanal. Escort was provided by task forces that included the carriers *Hornet* and *Wasp* and the battleship *North Carolina*. On the fifteenth, within a ten-minute period, the *Wasp*, *North Carolina*, and destroyer *O'Brien* were torpedoed by enemy submarines. Refueling her planes at the time, the *Wasp* was soon enveloped in flames. In spite of heroic efforts by her crew, heavy explosions forced her abandonment, and she was finally sunk by an American destroyer.

In October the Japanese decided to mount a major assault on Henderson Field. Assembling the largest force since Midway, they opened the Battle of Santa Cruz with a battleship bombardment of the U.S.-held airfield. Only one plane escaped damage. The bombardment was continued over the next several days by cruisers and destroyers. Reinforcement aircraft and fuel were flown in from Espíritu Santo.

The Japanese supporting force, with the carriers *Shokaku*, *Zuikaku*, and *Zuiho*, was opposed by Task Force 16, with the *Enterprise*, and Task Force 17, with the *Hornet*. In the ensuing exchange of raids, the *Shokaku* and *Zuiho* were damaged, as was the *Enterprise*. The *Hornet* was so heavily damaged that she was abandoned and sunk by American destroyers.

Many of the Japanese attempts to reinforce their garrisons in the Solomons were conducted at night. Hidden during the day, the reinforcing ships made high-speed runs through New Georgia Sound ("The Slot") to the landing beaches and attempted to put their troops ashore during darkness.

PBY Catalinas ranged far and wide through the Solomons, scouting movements of Japanese forces and attacking any and all targets found. Since the Japanese operated predominately at night, so did the PBYs. Equipped with radar and painted dull black to reduce their visibility, these planes and their aggressive crews came to be known as the Black Cats.

Unlike squadrons involved in antisubmarine warfare in the Atlantic, Pacific patrol squadrons were engaged primarily in anti-

During the Pacific war consolidated PBY Catalinas painted flat black prowled the South Pacific at night, sinking thousands of tons of enemy shipping. They were known as the Black Cats. (U.S. Navy)

shipping, search and rescue, and scouting operations for the fleet. They attacked everything from battleships and carriers to coastal shipping and fishing boats. In the search-and-rescue role, they excelled in "Dumbo" rescue missions in support of carrier strikes and later in support of the Army Air Forces' strategic bombing campaign against Japan. Seldom operating from fixed bases, patrol squadrons and their supporting tenders moved across the Pacific with the fleet, operating from sheltered water or advanced air-strips.

The Allied conquest of the Solomons proceeded in leapfrog fashion in a series of operations that often bypassed Japanese-held islands. By capturing airfields and maintaining control of the air, the Allies effectively isolated the bypassed areas.

Following heavy losses on both sides in 1942, it was almost a year before carrier forces again became available in strength. By the fall of 1943, the United States was ready to take the offensive. The first of the *Essex-* and *Independence*-class carriers had joined the fleet, and a stream of new pilots, aircrewmen, planes, and support came with them.

The Japanese, entrenched in the Gilbert and Marshall islands, were among the first to experience the destructive capabilities of this strengthened force. The subsequent action moved westward into the Carolines and northwestward toward the Marianas, Okinawa, and Japan itself.

The Marianas campaign posed a major threat to Japan. Operations had now moved to the inner defense area, and the Japanese fleet could no longer forego a reaction. The Japanese met the challenge with their own fast-carrier task force. During the ensuing Battle of the Philippine Sea, Japanese losses were staggering. In what later came to be called the Marianas Turkey Shoot, the Japanese lost 402 planes, the Americans 17. Japanese carrier aviation was essentially finished as a fighting force.

Subsequent operations in the Marianas campaign consisted of air support for the landings on Guam and Tinian and carrier strikes

Task groups built around aircraft carriers pursued the retreating Japanese across the Pacific. These ships of Task Group 50.3 have just completed strikes against the enemy in the Philippines. (U.S. Navy)

and photoreconnaissance against the Bonin and western Caroline islands.

By the time the Philippines campaign began in October 1944, U.S. naval aviation had developed the most powerful naval air striking force in history.

Initial landings commenced on the island of Leyte on 17 October. Air support was provided by sixteen escort carriers screened by nine destroyers and eleven destroyer escorts. The fast-carrier task groups alternated in supporting the troops ashore.

The Battle of Leyte Gulf developed as the Japanese, reacting to the serious threat of the landings, prepared a three-pronged operation to drive U.S. forces from the Philippines. It was a disaster for the Japanese.

THE END IN SIGHT

The occupation of the Marianas had given the United States airfields from which the strategic mining and bombing of Japan could be conducted. Iwo Jima, which was taken in March 1944, provided a base for long-range fighter escorts for the bombers and denied the Japanese the benefit of early warning. The lifelines of the Japanese were now under continuous attack from both submarines and aircraft, but her field armies were still intact. Japan was not yet at the point of surrender.

One last base was needed to support the projected invasion of the Japanese home islands. That base had to be sufficiently large to support the advanced airfields, supply depots, and fleet-support facilities needed for such a gigantic undertaking. Okinawa filled the bill.

After the war in Europe ended, the master plan envisioned transfer of vast Allied forces thus released to the Pacific theater for the final operation against Japan. It was against this background that the Okinawa campaign was planned. Operations in support of the landings proceeded in routine fashion until 6 April 1945. The day dawned with a call to general quarters to repel air attack. Before the day was over, more than four hundred suicide flights were launched against the American fleet; the kamikazes flew everything from trainers to the latest fighters and bombers.

The flying bomb, flown by a man on a one-way flight, made its appearance over the fleet on 12 April. It was dubbed *Baka*, Japanese for "fool," by the Americans. A previously damaged U.S. destroyer was hit and sunk by a Baka on the twelfth.

Between 6 April and 28 May, the crucial period of the Okinawa campaign, seven massed suicide raids were conducted by the Japanese. The grand total of enemy aircraft destroyed during this period was 3,594, of which Task Force 58 accounted for 2,259. American combat and operational losses were 880, including 266 planes lost on damaged carriers. Of 253 Allied ships sunk or damaged, suicide attacks were responsible for 189. Task Force 58 did not lose any carriers, and valiant damage control efforts by their crews saved the *Franklin* and *Bunker Hill*. The greatest suicide offensive in history had failed. The fleet had come to stay.

Marine Air Group 31 moved into Yontan Airfield on 7 April and immediately launched combat air patrols. Marine Air Group 33 arrived two days later, and within six days both groups were engaged in close air support. Kadena Airfield became operational soon thereafter. On May 13 army P-47s, flown in from Saipan, augmented the marine effort. On 16 May the first distant fighter sweeps over Kyushu were flown.

Following a period of rest and recreation in Leyte Gulf, fast carrier Task Force 38, under Vice Admiral John S. McCain, sortied on 1 July. Its composition at that time was nine large and six light carriers, nine battleships, nineteen cruisers, and numerous destroyers. On the eighth it was joined by the Royal Navy's Pacific Fleet, composed of four fast carriers, one battleship, six cruisers, and eighteen destroyers.

Initial strikes were made against military and industrial targets in the greater Tokyo area. Following four days of bad weather, the task force moved north and attacked targets and shipping in and around northern Honshu and Hokkaido. Again moving south, strikes were conducted against the Kure Naval Base and targets of opportunity in the Inland Sea. Due to unfavorable weather in the Tokyo area, the strikes were shifted to the west coast of Honshu, where industrial targets in Nagoya and Maizuru were bombed.

On 6 August a lone B-29 dropped a single bomb that destroyed

Navy pilots decimated Japanese air power in the Pacific. David McCampbell, the navy's top ace, had thirty-four enemy aircraft to his credit. (U.S. Navy)

the heart of the city of Hiroshima, signaling that the end of the war was near. On the ninth Nagasaki was destroyed by the second atomic bomb.

Conventional air strikes on Tokyo were resumed on the thirteenth, and the task force refueled the next day. Between the first and second strike on the fifteenth, word of the Japanese surrender was received. The navy's air war had come to an end.

The mighty effort to forge the navy's air weapon had produced results far beyond the wildest dreams of the prewar planners. Between 1 July 1940 and the end of the war, over eighty-three thousand planes had been built for the navy. A worldwide airline, the Naval Air Transport Service, had been established. Over one hundred aircraft carriers of all types had been constructed, as well as seaplane tenders and other aviation support ships. At the end of the war, almost four hundred thirty-one thousand trained navy and marine corps officers and enlisted personnel were serving in aviation.

In protecting merchant convoys carrying the commerce of war throughout the world and in spearheading the massive drive to reconquer the western Pacific, naval aviators produced an impressive record of success. Planes and their crews sank 174 Japanese warships, including 13 submarines and 447 large merchant vessels, and assisted in the destruction of many more ships and numerous small craft. In the Atlantic, navy planes and blimps destroyed 63 German submarines and assisted in the destruction of an additional 20.

PEACE AND PROGRESS

With the coming of peace, naval aviation was reduced to about a tenth of its wartime strength. Ships, stations, and units were decommissioned on a wholesale basis. But some of the mistakes of the previous world war demobilization were not repeated. Ships and planes were "mothballed" so that they could be used at a later date should the need arise. A strong naval air reserve was created, supported by naval reserve air stations throughout the country.

The fruits of wartime research soon appeared as the first armored-deck carriers, the *Midway* class, joined the fleet. Air groups began to receive aircraft designed during the war but built too late for combat operations. The F8F Bearcat, AD Skyraider, FJ Fury, FH Phantom, and F9F Panther became the mainstays of the carrier air groups. The P2V Neptune and the P4M Mercator joined the fleet in the late 1940s.

Helicopters, which had become available in small numbers during the war, now began to prove their value and soon replaced observation planes on the battleships and cruisers. Carriers carried one or more of these aircraft, principally for close-in search-and-rescue operations.

New developments in lighter-than-air aviation included new, larger envelopes for some of the wartime K- and M-class blimps and introduction of the postwar N class.

Weapons development included advanced aerial rockets and drone aircraft for various uses. A comprehensive series of atomic bomb tests was conducted at Bikini Atoll in the Pacific. The damage to a large number of warships was evaluated in these tests, including that to the fast carrier *Saratoga* and the light carrier *Independence*. The lessons learned, including those gleaned from damage-control data, were applied to the design of subsequent classes of warships.

In March 1946 the chief of the Bureau of Aeronautics proposed to the commanding general of the Army Air Forces that a joint project be established to develop an earth satellite. Eventually both the army and navy established satellite programs.

Project Skyhook, strongly supported by naval aviation, developed and launched a number of large balloons used for research at the outer limits of the earth's atmosphere. Weather squadrons were supported in both the Atlantic and Pacific, to detect, track, and report hurricanes and typhoons.

Operation High Jump, the largest single Antarctic exploring expedition, was carried out in 1946–47. Shore-based aircraft flown from the carrier *Philippine Sea* and PBM Mariner flying boats from several seaplane tenders photomapped 1.5 million square miles of the continent and 5,500 miles of coastline.

The Sixth Fleet in the Mediterranean and the Seventh Fleet in the western Pacific were established to support U.S. interests in their respective areas. As a consequence of Communist expansion efforts in the Cold War, American forces were once again committed overseas.

A series of international crises in various parts of the world sorely taxed the overextended forces of naval aviation. It seemed as if the numbers of carriers and squadrons were in some inverse relation to the number of aircraft needed to handle crises.

AGAIN THE CALL

On 25 June 1950 forces of North Korea invaded the Republic of South Korea. President Truman responded on 30 June by committing U.S. forces in support of a United Nations resolution calling for a North Korean withdrawal and restoration of the integrity of South Korea. On 3 July U.S. Navy and Royal Navy carriers went into action against the North Koreans. The war in Korea was a war of limited objective—to enforce the United Nations resolution. Though the bulk of engaged United Nations forces came from the United States, numerous allied nations provided units, most of which received logistic support from the United States.

During combat operations in Korea, the attack carriers generally operated under Task Force 77 in the Sea of Japan, the light and escort carriers under Task Force 95 in the Yellow Sea. American, British, and Australian carriers rotated duty in the Yellow Sea. Patrol squadrons operated under Fleet Air Wing 6. Farther south, planes of Fleet Air Wing 1 patrolled the coast of China. Squadrons of the First Marine Air Wing operated from airfields in Japan and Korea as well as from the escort carriers of Task Force 95.

The Korean War saw navy jet aircraft used in combat for the first time. F9F Panthers heavily damaged military and industrial

Navy jets were first put to the test of combat in Korea. Here Grumman F9F-5 Panthers streak by the USS *Princeton* (CVS 37) in Korean waters in 1951. (U.S. Navy)

targets in Pyongyang, capitol of North Korea, on 3–4 July 1950. Naval aviators downed their first MiGs during strikes on the bridges over the Yalu River.

For the most part, however, enemy air opposition during the course of the Korean War was generally light to nonexistent in the areas where most naval air operations were conducted. Seventeen enemy aircraft were downed by navy pilots, including five by Lieutenant G. B. Bordelon, the navy's first "night ace." Marines accounted for thirty-five enemy aircraft, including six downed by Major J. F. Bolt, the marines' only Korean War ace.

Patrol squadrons flew routine search and reconnaissance missions as well as mine spotting and surveillance of merchant shipping and fishing activities. For the first time, land-based patrol planes were used in greater numbers than flying boats, a trend that would continue until the flying boat disappeared from the naval aircraft inventory in the late sixties.

For the most part, the mission of the attack-carrier air groups remained interdiction and occasionally close air support for the remainder of the war. However, several missions of a different nature served to break the monotony.

In May 1951 large air-dropped antishipping torpedoes were used for the last time. The Hwachon Reservoir dam was situated where it controlled the water level of the Pukhan and Han rivers. The Communists had used the ability to vary the water levels in these rivers to their advantage several times. B-29 strikes against the dam had not been successful.

On 30 April VA-195, flying from the *Princeton*, was given the mission of breaching the dam. A preliminary raid provided flak suppression and put one hole in the dam. On 1 May eight AD Skyraiders armed with torpedoes made their attack. Six torpedoes ran straight and hot. The center sluice gate was completely destroyed, and a large hole was opened in another. The waters of the reservoir were released.

To give impetus to the stalled truce talks, a coordinated navy–air force operation against the North Korean hydroelectric system was authorized. Thirteen major North Korean power plants were attacked during a two-day series of strikes beginning on June 23.

The major strike of this operation was that flown against the Suiho power plant, fourth largest in the world. It was situated on the Korean side of the Yalu River. ADs and F9Fs from the *Boxer*, *Princeton*, and *Philippine Sea* flew the mission. High cover was provided by air force F-86 Sabers. The navy strike was completely successful. As navy planes pulled off target, air force F-84 Thunderjets pulverized what remained of the Suiho installation. Concurrently, twelve other North Korean power plants were attacked.

The helicopter came into its own during the Korean War. It was used in a wide variety of missions, ranging from utility to medical evacuation. It was largely through the use of the helicopter in combat search and rescue that a number of aviators shot down behind enemy lines were returned to their units.

During the Korean War, navy and marine corps planes flew 276,000 sorties, dropped 177,000 tons of bombs, and fired 272,000 rockets. This effort was only 7,000 sorties less than those flown by naval aviation in all theaters during World War II.

FORGING AHEAD

The decade following the Korean War was to see significant improvements in the material aspects of naval aviation. Conversion of *Essex*- and *Midway*-class carriers to angle-deck configuration was completed. Steam catapults and the mirror landing system became standard on attack carriers. New, larger attack-carrier classes, including the first nuclear-powered carrier, the *Enterprise*, joined the fleet.

Airborne-early-warning barriers were established to provide radar coverage to the Atlantic and Pacific approaches to the United States. They served as seaward extensions to the "Dew Line" radar network across arctic North America. WV Warning Stars flew millions of miles on the barrier patrols during the nine years they were in operation.

In preparation for the International Geophysical Year (1957–58), the navy, with the assistance of the other services, mounted the most extensive Antarctic exploring expedition ever conducted, Operation Deep Freeze. The first landing ever at the geographic

South Pole was made by an R4D Dakota from Air Development Squadron 6 in October 1956. This station has been continuously manned and supplied by naval aviation ever since. Other isolated stations were established along the coast of the continent and inland at elevated locations on the polar plateau.

Antisubmarine warfare was established as the primary mission of patrol aviation in the early 1950s. Patrol squadrons were equipped with various models of the P2V Neptune and the P5M Marlin flying boat. A number of *Essex*-class carriers were converted to antisubmarine duty.

The S2F Tracker, combining search-and-attack capabilities in one carrier-based aircraft, made its debut. Antisubmarine helicopter squadrons were equipped with the HSS Seabat until this was replaced with the SH-3 Sea King in the early 1960s. The carrier airborne-early-warning mission was flown by the AD-5W, a modified Skyraider.

With the introduction of more powerful and reliable jet engines, a new family of fighter and attack aircraft emerged that brought a number of international performance records to the United States.

As had been the case following World War II, naval forces supported by naval aviation were involved in operations ranging from a show of force to providing emergency relief to victims of natural disasters. During the Berlin crisis of 1961, five patrol and thirteen carrier antisubmarine squadrons of the naval reserve were called to active duty.

The Cuban Missile Crisis of 1962 resulted in a naval quarantine of Cuba. Extensive naval air operations involved search, tracking, and reconnaissance. Low-level missions flown by Light Photo Squadron 62 produced proof of Soviet adventurism, which was used by the United States before the United Nations. As a result of American resolve, the Soviet Union withdrew their offending missiles and bombers from Cuba.

Culminating forty-four years of operations, the last remaining units of the navy's lighter-than-air branch, ZP-1 and ZP-3, were disestablished at the naval air station at Lakehurst, New Jersey, in October 1961.

In the early 1960s, a new dream captured the imagination of

the American public. NASA set a goal of putting a man on the moon before the decade ended. Four of the first seven astronauts were naval aviators. During the Mercury program, Lieutenant Commander Alan Shepard became America's first man in space with his suborbital flight in May 1961. Marine Lieutenant Colonel John Glenn became the first American to orbit the earth in February 1962. The navy provided the means of recovery for both astronauts and spacecraft. Hundreds of ships and aircraft were engaged in these recovery operations. The follow-on Apollo program resulted in the first landing on the moon on 20 July 1969.

VIETNAM

On 2 August 1964 the destroyer *Maddox*, on an intelligence-gathering patrol, reported an attack by three North Vietnamese torpedo boats. A flight of F-8 Crusaders from the *Ticonderoga*, at the time airborne on a training mission, were diverted and sank one of the retiring torpedo boats. On the night of 4 August the *Maddox*, now accompanied by the *Turner Joy*, was purportedly again attacked by North Vietnamese torpedo boats. Following these two incidents, President Lyndon Johnson decided to take retaliatory action.

The *Ticonderoga* and *Constellation* launched strikes on 5 August that resulted in the destruction of more than one half of the North Vietnamese torpedo boat force.

Congress passed the Tonkin Gulf Resolution on 10 August 1964. Carriers remained on the alert in the South China Sea, prepared for any emergency. Patrol aircraft maintained surveillance over the area.

In response to further enemy attacks in March 1965, thirty-five hundred marines landed at Danang. To bring the Communists to the negotiating table, a bombing campaign, Rolling Thunder, was started. The air force and navy conducted strike operations progressing north from the demilitarized zone between the two Vietnams to the suburbs of Hanoi. As the bomb line moved closer to Hanoi, it was hoped that the North Vietnamese would be encouraged to seek peace.

Yankee Station was established in the Gulf of Tonkin to serve

Briefings finished, pilots man their planes aboard the USS *Coral Sea* for a strike against the enemy in Vietnam, 1965. (U.S. Navy)

as a central point for operations against North Vietnam. This was followed by the establishment of Dixie Station in the South China Sea as a central location for carrier operations in support of South Vietnam.

In 1965 the marines established a short airfield for tactical support at Chu Lai. It was initially equipped with arresting gear and later with a catapult. Marine squadrons operating in support of the First Marine Division used this field.

Activity in South Vietnam was relatively quiet during the latter part of 1967 and January 1968. However, reconnaissance flights during the Christmas truce indicated that all lines of communication between North and South Vietnam were choked with supplies. Meanwhile, the Communists were at the peace table.

In the spring of 1968 bombing was restricted to the area south of 20 degrees north latitude, and on 1 November President Johnson ordered a halt to all bombing of North Vietnam. As Rolling Thunder ended, air operations concentrated on interdiction of the infiltration routes through Laos. Popularly called the Ho Chi Minh Trail, this route now bore the brunt of the bombing effort.

Navy squadrons also made significant contributions in the Mekong River delta region of South Vietnam. Watercraft were the principal means of transport in this area, and the vital support for this traffic was largely provided by light helicopter and fixed-wing attack squadrons.

Several new aircraft entered the inventory during the Vietnam conflict. The A-7 Corsair II attack aircraft replaced the A-1 Skyraider on the carriers. The EA-6 Prowler, E-2 Hawkeye, and C-2 Greyhound brought a new dimension to each of their specialized missions. By the end of the war, all patrol squadrons were flying the P-3 Orion.

The year 1967 saw the end of an era as the last operational mission by a navy flying boat was flown by an SP-5B from Patrol Squadron 40. For much of their more than fifty years of operations, flying boats had been the mainstay of patrol aviation.

In the rotary-wing community, the H-1 Huey (or Cobra), H-2 Seasprite, H-46 Sea Knight, and H-53 Sea Stallion entered the inventory. The SH-2 fulfilled the antisubmarine helicopter mission on smaller ships and was designated the light airborne multipurpose system (LAMPS I).

Enemy air activity over North Vietnam was sporadic, but in May 1972, during intense fighting, navy fliers downed sixteen MiGs. From these battles Lieutenant Randy Cunningham and his radar

Naval aviators were involved in a different kind of war in Vietnam. This Bell UH-1B Huey gunship provides air cover for river patrol operations in the Mekong Delta. (U.S. Navy)

intercept officer, Lieutenant (j.g.) Willie Driscoll, flying an F-4 Phantom, emerged as the navy's only Vietnam Aces as well as its only team Aces.

On 8 May 1972 President Nixon ordered the mining of Haiphong and other North Vietnamese harbors. He also ordered the air war against North Vietnam resumed with maximum effort against lines of communication to the south. The campaign produced results, and the flow of supplies through Haiphong effectively stopped.

The ongoing peace negotiations resulted in a 23 October decision again to halt the naval and air operations north of 20 degrees north latitude. The Communists took advantage of the bombing halt to resupply their forces and to reinforce their antiaircraft and fighter forces with supplies brought in from China and the Soviet Union, the railroads from China again being open.

The United States, having once again given up the fruits of a successful air campaign, provided a sanctuary for a Communist buildup in hopes of getting serious negotiations going. Recognizing the futility of the bombing halt, President Nixon ordered the bombing resumed.

Operation Linebacker II was carried out between the eighteenth and twenty-ninth of December, with a thirty-six-hour break at Christmas. On the twenty-sixth, navy airfield suppression strikes prevented enemy fighter opposition from intercepting the complex attack mounted against Hanoi and Haiphong when bombing was resumed after the Christmas break.

Negotiations resumed at the Paris peace talks on 6 January 1973. On 23 January a cease fire was announced. The *Enterprise* aircraft flew the last mission of the war on 27 January. On 12 February the first prisoners of war (POWs) released by the North Vietnamese arrived in the Philippines. By 29 March 566 American POWs had been returned. Over two thousand American airmen had been killed and another fourteen hundred were missing. Over 7.4 million tons of bombs had been dropped on Indochina since 1965.

STILL ON STATION

With the exception of the *Lexington*, reclassified as a training carrier, the remaining *Essex*-class carriers were decommissioned by the mid-seventies. The S-3 Viking entered the inventory, and both fixed-wing and rotary-wing antisubmarine aircraft were added to the attack-carrier air wings.

In 1973 the *Midway* was homeported in Yokosuka, Japan, thus becoming the first U.S. carrier to be based overseas. The nuclear-powered carriers *Nimitz* and *Eisenhower* were commissioned in the late seventies. Two new carriers, the *George Washington* and the *Abraham Lincoln*, were authorized in 1981, and the *Theodore Roosevelt*, whose keel had been laid that same year, was completed in 1984 and will be commissioned in 1986.

The AV-8 V/STOL attack aircraft was assigned to the marine corps, and the F-14 Tomcat, equipped with the long-range Phoenix missile, entered the inventory. This combination brought to the fleet an air-defense capability far beyond that previously available.

Naval aviation has kept a continuous presence in the Mediterranean since World War II. Here an F-4J Phantom II fighter lands aboard the USS *Franklin D. Roosevelt* (CVA 42). (U.S. Navy)

The emphasis in space had shifted by 1973 from the exploration of the moon to scientific experimentation in earth orbit. The Skylab project found navy and marine corps astronauts engaged in various activities aloft.

The recent era has seen expanded opportunities for women in naval aviation. Although women had been previously designated as aircrew, in 1974 the first women qualified as flight surgeons, and on 22 February Lieutenant (j.g.) Barbara Ann Allen became the first woman to be designated a naval aviator.

During the 1973 Arab-Israeli War, the *Franklin D. Roosevelt, Kennedy, Independence, Guadalcanal,* and *Iwo Jima* were deployed to the Mediterranean. Fifty A-4 Skyhawks were ferried to Israel using en route air-to-air refueling, staging through the Azores and the *Roosevelt.* United States forces were placed on a worldwide increased-alert status in response to the possibility of unilateral Soviet intervention in the conflict. Meanwhile, the carriers in the

Mediterranean provided protection for the massive airlift of material from the United States to Israel. The *Hancock* deployed to the Indian Ocean for possible contingency operations resulting from the Arab oil embargo. The end of the conflict once again saw the navy's aerial minesweeping capability utilized during the clearance of the Suez Canal and its approaches.

Beginning in 1979, naval aircraft participated in operations to assist the "boat people," refugees from Vietnam fleeing to freedom across the South China Sea or the Gulf of Siam in small, overcrowded, unseaworthy boats.

Since the Iranian takeover of the United States embassy in Teheran in November 1979 and the Soviet invasion of Afghanistan the following month, carrier battle groups have been deployed to the Indian Ocean for possible contingency operations.

In April 1980 eight RH-53D Sea Stallions, flying from the *Nimitz* with marine corps crews, participated in a joint operation to rescue the fifty-two hostages held in the American embassy in Teheran. Mechanical difficulties forced cancellation of the mission, and several personnel were lost in a tragic mishap during the withdrawal of the rescue force.

In August 1982, during missile-firing exercises over international waters in the Gulf of Sidra, off the coast of Libya, naval aircraft were harassed by Libyan fighters. Two VF-41 F-14 Tomcats from the *Nimitz* were fired upon by the Libyan aircraft. They missed. The Libyan pilots' Soviet-built SU-22s were downed by missiles from the Tomcats. Following this incident, harassment stopped, and the exercises were successfully concluded.

During operations to prevent Communist expansion into the eastern Caribbean, combat amphibious assault operations on Grenada in October 1983 were supported by navy and marine air operations from the *Independence* and *Guam*. Surveillance was provided by Atlantic Fleet patrol squadrons, and several reserve transport squadrons provided combat support.

In December 1983, in retaliation for attacks on reconnaissance missions over Lebanon, the *John F. Kennedy* and *Independence* launched a coordinated strike against Syrian antiaircraft positions. Several aircraft were lost to the intense ground fire. Lieutenant

Robert O. Goodman ejected from his damaged A-6 Intruder, and was made a POW. He was later released by the Syrians.

The era of the navy's enlisted pilots came to an end in 1981. From the first class of enlisted men selected for flight training in 1916 to the last of the line, Master Chief Petty Officer R. K. Jones, the Silver Eagles served their country with honor. Never a large group, the NAPs experienced many changes in their program over their years of service. Their last flight training class graduated in 1947.

New aircraft reaching operational status in the 1980s include the free world's largest helicopter, the CH-53E Super Stallion, the SH-60B LAMPS III, and the F/A-18 Hornet. Prototypes of the

Captain Bruce McCandless becomes the first man to walk untethered in space during Space Shuttle *Challenger*'s mission 41-B in February 1984. (National Aeronautics and Space Administration)

MH-53E and the P-3C Update III are flying, and the F-14D and A-6E Update are entering full-scale development. Advanced concepts now being explored hold great promise for the future. New weapons under development include air-to-air and air-to-ground missiles, retarded bombs, and an advanced lightweight torpedo for antisubmarine operations.

After six years without manned space activities, the Space Shuttle Program was inaugurated in April 1981 with the launch of the Space Shuttle *Columbia*, the first reusable space vehicle. Carrying an all-navy crew, the *Columbia* was launched from Cape Canaveral, Florida, and was recovered at Edwards Air Force Base, California, two days and thirty-six earth orbits later. In February 1984 naval aviator Captain Bruce McCandless made the first untethered space walk from the Space Shuttle *Challenger* using a nitrogen-propelled manned maneuvering unit. Space Shuttle flights, many with navy crewmembers, are now conducted on a regularly scheduled basis.

Today naval aviation personnel around the world aboard carriers at sea and at strategic locations ashore maintain their solemn vigil in support of U.S. and allied interests. At home new weapons systems and equipment are constantly being tested and developed to keep pace with rapidly changing technology. Meanwhile, new naval aviators and support personnel, the lifeblood of naval aviation, are being trained and integrated into operational squadrons. The legacy continues.

2

Wings:
The Challenge

Revised by Commander Van N. Stewart, USN

What does it take for someone to become a navy pilot or a naval flight officer (NFO)? He or she must meet the physical requirements, of course, and have a solid academic background. But there is much more to becoming an aviator than that. It requires a combination of talents that many people possess but few are challenged to use to full measure. Some spend their lives avoiding challenge and never know the extent of their real capabilities. Others choose a goal, take a deep breath, and pursue it.

The key to success in naval aviation is personal commitment, and each individual must decide whether he is willing to make that commitment. Navy wings are not for everyone. In the final analysis, they are for those who will accept the challenge—those who will settle for nothing less.

Navy undergraduate flight training is conducted by the Naval Air Training Command and leads to a designation as a navy pilot or NFO. There are three ways of entering undergraduate flight

The Challenge (McDonnell Douglas)

training: directly from civilian life through the aviation officer candidate (AOC) program; as an officer already commissioned through another program such as the Naval Academy or NROTC; as an active-duty enlisted member serving in pay grades E-5 through E-7 (the flying limited duty officer, or FLDO, program).

THE AVIATION OFFICER CANDIDATE

The AOC program is open to male and female citizens of the United States who are between the ages of nineteen and twenty-nine and who have graduated from an accredited college or university with a baccalaureate degree. Candidates are commissioned ensigns in the U.S. Naval Reserve after successful completion of a fifteen-week course at the AOC school located at the naval air station in Pensacola, Florida. From this point they continue training until they earn their wings—approximately eighteen months

Aviation officer candidates parade at Pensacola on graduation day prior to receiving commissions as ensigns. (U.S. Navy)

for pilots, twelve months for NFOs. To apply for the AOC program one should contact any navy recruiting station. They are listed in telephone directory white or blue pages under U.S. Government.

Officers who have already earned their commissions from other sources follow a slightly different path initially. These students receive six weeks of aviation indoctrination at the Aviation School Command at Pensacola before beginning pilot or NFO training.

The FLDO program provides an opportunity for top-notch active-duty enlisted members serving in pay grades E-5 through E-7 with at least four years of service to earn commissions and their navy wings of gold. Applicants must be no more than thirty years old and have at least sixty college credits or a service-accepted equivalent. For additional information about aptitude and physical qualifications, consult the career information officer or the current Navy Military Personnel Command (NMPC) notice (NAVMILPERS-COMNOTE 1131).

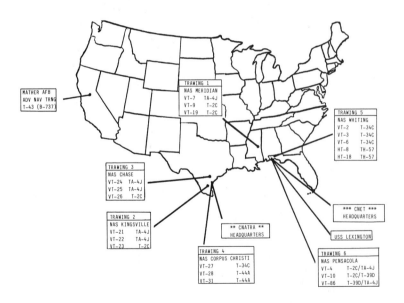

Fig. 2-1. Location of Training Squadrons

GENERAL INFORMATION

Undergraduate training for both pilots and NFOs is conducted at a number of locations through six training wings, eighteen fixed-wing training squadrons, two helicopter training squadrons, and the training aircraft carrier USS *Lexington*.

Training squadrons prepare pilots and NFOs to operate aircraft and their equipment. After basic training, students separate into pipelines, where training is more advanced and oriented toward the specialized missions of naval aviation. The twenty training squadrons, their locations, their missions, and the type of aircraft they fly are listed below:

Squadron	Location	Type of Training	Aircraft
VT-2	NAS Whiting, Milton, FL	Primary/Intermediate	T-34C
VT-3	NAS Whiting, Milton, FL	Primary/Intermediate	T-34C
VT-6	NAS Whiting, Milton, FL	Primary/Intermediate	T-34C
HT-8	NAS Whiting, Milton, FL	Basic Helo	TH-57
HT-18	NAS Whiting, Milton, FL	Advanced Helo	TH-57
VT-7	NAS Meridian, MS	Advanced Strike	TA-4
VT-9	NAS Meridian, MS	Intermediate Strike	T2-C
VT-19	NAS Meridian, MS	Intermediate Strike	T2-C
VT-4	NAS Pensacola, FL	Intermediate/Advanced Strike	T2-C/TA-4
VT-10	NAS Pensacola, FL	Basic/Intermediate NFO	T2B/T-47
VT-21	NAS Kingsville, TX	Advanced Strike	TA-4
VT-22	NAS Kingsville, TX	Advanced Strike	TA-4
VT-23	NAS Kingsville, TX	Intermediate Strike	T2-C
VT-24	NAS Chase Field, Beeville, TX	Advanced Strike	TA-4
VT-25	NAS Chase Field, Beeville, TX	Advanced Strike	TA-4
VT-26	NAS Chase Field, Beeville, TX	Intermediate Strike	T2-C
VT-27	NAS Corpus Christi, TX	Primary/Intermediate	T-34C
VT-28	NAS Corpus Christi, TX	Advanced Maritime	T-44
VT-31	NAS Corpus Christi, TX	Advanced Maritime	T-44
VT-86	NAS Pensacola, FL	Advanced NFO	T-47/TA-4J

Note: Advanced navigation for NFOs is taught at Mather Air Force Base, California, in a course conducted jointly by the navy and air force.

Naval aviation training for officers begins at the naval air station in Pensacola, Florida, which is often referred to as the Annapolis of the air. Before the student can get into the air he must undergo a period of intensive indoctrination and ground school. This includes military, academic, and physical fitness/survival training. The fifteen-week course for AOCs learning to be officers is conducted separately from the five-week course for those already commissioned. The course for the former group includes rigorous military training and indoctrination into military life. Students learn the manual of arms, military etiquette and custom, formation, and standard military procedures.

All students, officers and candidates alike, undergo academic training to prepare them for flight training. The course of study includes math, physics, engineering, aerodynamics, and aviation physiology.

The physical fitness/survival part of the course is also mandatory

There is much to learn before climbing into a cockpit. Here preflight students give rapt attention during a class in aerodynamics. (U.S. Navy)

for all students. A rigorous conditioning course includes pro-
grammed exercise, an obstacle course, and a cross-country run.
Because much of a pilot or NFO's time is spent over water, a
considerable amount of training is devoted to swimming and water

Students must know how to enter the water safely and disengage themselves from a parachute harness in the event of an emergency. (U.S.
Navy)

survival. Students must master the basic swimming strokes and swim a mile fully clothed. They are also dunked while strapped into a simulated cockpit known as the Dilbert Dunker and dragged through the water in a parachute harness to learn how to extricate themselves after ejecting from an aircraft over water. Students are taught to survive in the wild and live off the land if necessary. They learn the difference between harmless and dangerous animals and vegetation and must practice their skills in realistic survival training exercises in the field.

Because of the rigorous nature of the physical fitness/survival program, students should be in good physical condition when they arrive at Pensacola.

At the end of fifteen weeks, AOCs receive their commissions as ensigns in an impressive commissioning ceremony. As full-fledged naval officers they take their places beside those commissioned through other programs and begin in earnest their progress toward navy wings. Those who become pilots take one route, while those destined to become NFOs take another.

PILOT TRAINING

All student naval aviators begin the actual business of flight in primary training, conducted mostly at Whiting Field, near Pensacola, in VT-2, VT-3, or VT-6 (training squadrons). Some students will undergo primary training at Corpus Christi, Texas, in VT-27. Regardless of location or squadron, the training aircraft will be the turboprop Beech T-34C Mentor. Before strapping in and taking to the air, however, students must undergo familiarization training in a mock-up of the aircraft, where they learn the exact position of the flight controls and operating switches as well as engine and flight instruments. Then they are ready to fly.

A flight instructor is assigned to four or five students and stays with them through the several weeks of primary training, giving encouragement, advice, and criticism as necessary. The instructor has a large stake in his students' success. Their progress reflects his competence and effort, and he takes understandable pride in their achievement. After approximately thirteen flights and many hours of consultation and briefings, the student is ready to solo.

This is the first big test, the moment when students lift the plane off the ground by themselves—and bring it back safely. They will have many memorable experiences thereafter, but none will equal the thrill of that first solo.

After more solo flights and additional instruction in formation and basic instrument flying, primary training ends. At this point the student aviators separate into three pipelines, strike, maritime, and rotary wing. Strike training leads to assignment to carrier-based jet aircraft (A-6, A-7, F-14, F-18, S-3, etc.). The maritime pipeline prepares the student for large, multiengine, turboprop patrol, reconnaissance and support aircraft (P-3, C-130, etc.). Rotary-wing training leads to helicopters (SH-2, SH-3, RH-53, CH-53, CH-46, etc.). A few pilots destined to fly turboprop E-2 early-warning and C-2 logistics aircraft follow a unique path through parts of both the maritime and strike syllabi. Those assigned to

The student pilot's first aircraft is the Beech T-34C Mentor. (U.S. Navy)

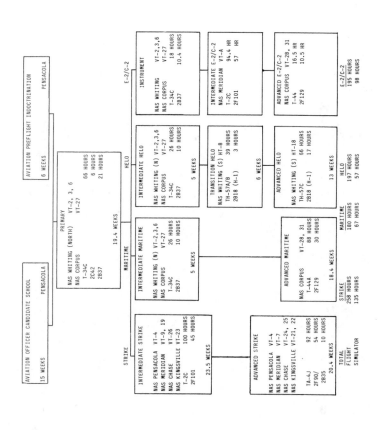

Fig. 2-2. Pilot Training

the strike pipeline move on to one of four training air wings located in Pensacola; Meridian, Mississippi; and Kingsville or Beeville, Texas. Helicopter pilots remain in the Pensacola area at Whiting Field, while maritime pilots proceed to Training Air Wing 4 at Corpus Christi, Texas.

Maritime and helicopter intermediate training each take about five weeks. During this period the student in each pipeline accumulates twenty-six flying hours in fourteen flights devoted to more advanced instrument work, day- and nighttime visual navigation, and airways navigation.

Checkout in the ejection-seat trainer is a must. (U.S. Navy)

While the maritime and helo students are undergoing their short intermediate courses at Whiting or Corpus Christi, the student strike pilots are undergoing a similar but longer training cycle at one of the strike training wings at Meridian, Pensacola, Beeville, or Kingsville. They train in the T-2 Buckeye jet aircraft and receive approximately one hundred hours of flight instruction, both dual and solo.

Cruising at speeds in excess of 400 miles per hour at 30,000 feet becomes a matter of routine for the strike trainees. High- and low-altitude flying, high-speed formation, precision acrobatic flights, as well as visual and instrument navigation open new vistas to the student aviator.

Intermediate strike training also includes hours of individual classroom instruction in meteorology, communications, engineering, aerodynamics, and navigation, as well as demanding hours in the 2F-101 flight simulator, which allows each student to practice in detail the procedures and maneuvers required in flight.

A student pilot flying a Rockwell International T-2C Buckeye prepares to qualify aboard the training carrier *Lexington* (AVT 16). (Harry Gann)

The final intermediate training stages for the strike student are gunnery and carrier qualification aboard the training carrier *Lexington*. The gunnery syllabus consists of dual and solo instructional flights where the student must demonstrate his ability to maneuver his aircraft, in company with three others, into a firing position on a towed banner while flying at speeds in excess of 300 knots.

Carrier qualification is another special rite of passage for the student aviator. Ten field-practice periods lead to a shipboard-qualification flight. Students are given extensive lectures on the optical landing system by their landing signal officer (LSO). After two field-practice flights with an instructor they are on their own. After demonstrating their proficiency in the field carrier-landing pattern, students are ready to go aboard the *Lexington*, operating in the Gulf of Mexico. They are required to complete two touch-and-go landings, four arrested landings, and four catapult launches. Then, with a new sense of confidence, they move on to the final stage of training.

Under the single-basing strike training concept, students receive advanced instruction at the same base at which they completed the intermediate stage. The advanced program is divided into instruments, night flying, formation, air combat maneuver, weapons training, field carrier-landing practice, and carrier qualification in a new aircraft.

It is at this point that the student is introduced to the TA-4J Skyhawk swept-wing, transonic jet aircraft by way of the 2F-90 operational flight simulator. Through training in the simulator, he is familiarized with the cockpit of the new aircraft and able to practice the new procedures.

The flying picks up again with more instrument training, which develops the student's confidence in his ability to fly an operational aircraft in all weather conditions. Here he will refine and build on the instrument techniques begun earlier in the program. When this phase of training is complete, he will be qualified to fly under instrument flight rules (IFR) and will have demonstrated competence in airways navigation as well as instrument departures and approaches to landing, including ground-controlled approaches by radar. During the night-flying portion of the program, the student will adapt his skills to the night environment and become proficient in night formation, night navigation, and night approaches and landings.

Formation flying, in which military pilots and their aircraft combine their skills to become mutually supporting, offers a different kind of challenge. It facilitates better coordination of assets and permits concentration of firepower on a target. Student pilots learn to rendezvous with other aircraft, cruise in formation, lead the formation, conduct high-altitude maneuvers, and perform carrier-breakup techniques used to enter the traffic pattern at the ship.

One of the most strenuous, demanding, and stimulating portions of the course is air combat maneuver. The student is taught about relative motion and to deal with evasive hard turns and breaks, high rates of closure between aircraft, and attack and defensive maneuvers. Here he learns team tactics and some of the tricks of the trade, such as forcing an opponent to overshoot an approach.

In the weapons phase of training, conventional air-to-ground ordnance-delivery techniques are introduced. In bombing, the parameters of dive angle, line of flight, release points, target motion, and ballistics are all considered. Nuclear weapons delivery techniques and close-air-support tactics using rockets and gunfire are also taught during this phase.

Mission planning is a vital adjunct to flight. Here the student grasps the importance of precision navigation at high altitudes and high speeds as well as the criticality of high fuel consumption. He will learn cruise control for maximum range or endurance, chart reading, and cross-country and tactical planning based on weather and load and fuel factors. Emergency procedures, check-off lists, en-route check points, and approach at destination are all part of mission planning.

The technique of landing an aircraft on a carrier has changed considerably since World War II. The introduction of the optical landing system puts a greater share of the responsibility for a well-executed recovery on the pilot. The man with the paddles is gone and today's navy pilot relies on the "meatball," an amber spot of light that gives him precision glide path information all the way down. The LSO can still be of considerable assistance on a pitching deck, but the aircraft's line-up, angle of attack, and air speed are largely up to the pilot. The student will be closing the ship in his TA-4J along a 3.5-degree glide slope at about 125 knots. He flies his aircraft to keep the "meatball" centered between two rows of green lights. Before taking his aircraft aboard ship, the student must demonstrate his proficiency ashore during field carrier-landing practice. Even though he has done it during the intermediate stage, landing an operational aircraft on a ship gives the student a special feeling of satisfaction and prepares him for high-performance fleet aircraft.

The navy is now entering an exciting and innovative period in jet aviator training. To replace the aging T-2 and TA-4 trainers, which are fewer in numbers than before and expensive to operate, the navy asked industry to design a complete training system to include new aircraft, new simulators, and a ground-training program that would give the student the best possible instruction from start to finish. This resulted in a modern, efficient aircraft, the

The T-45 jet trainer will be phased into training programs during the late 1980s. (McDonnel Douglas)

T-45, and a balanced mix of training media, including computer-aided instruction, realistic simulators, and a computer-based training management system. This is the first complete flight training system to meet specific learning objectives. Known as T45TS, it is expected to be in use by 1990.

Advanced training for students in the maritime pipeline begins at the naval air station in Corpus Christi, Texas, and lasts about eighteen weeks. A student typically receives about eighty-eight hours of flight time in the twin-engine T-44 *Pegasus*, which belongs to VT-28 or VT-31.

Advanced maritime training begins with ground school. The course of instruction includes aviation safety, aircraft procedures, and familiarization with the local area. Following completion of classroom work, the student is assigned an instructor and is introduced to his new aircraft.

After the student becomes completely familiar with the peculiarities of a twin-engine airplane, including single-engine operation, the instrument phase of training begins. What the student learns here will be transferred later to a large, heavy, multiengine aircraft with a full crew. He will be taught preflight planning, instrument takeoffs and departures, airways procedures, advanced patterns, partial panel techniques, and approaches to destination, as well as a variety of emergency procedures. He will master the use of navigation aids, the omnirange, TACAN, the instrument landing system, and radar. At the end of this phase the student will be subjected to a comprehensive instrument flight check and, on successful completion, will receive his instrument rating. At this point he will be "turned loose" with another student of equal qualification, and the two will sharpen their skills for the operational responsibilities that ultimately follow. Night flying is also an important part of the syllabus. When the student has completed the maritime course he will be confident of his ability to prosecute any assigned mission in any weather, day or night.

Some of the most demanding tasks in naval aviation are to be found in large, multiengine aircraft. There is a special satisfaction, not unlike that experienced by the commanding officer (CO) of a ship, in directing the efforts of a crew of professionals to achieve a goal. The crew of a P-3 Orion, for example, spends long hours together in one of the most important and demanding missions of naval aviation—antisubmarine warfare. They become a finely honed team under the leadership of the patrol plane commander.

A limited number of students undergo a specially tailored fixed-wing program to prepare them to fly either the E-2C *Hawkeye* carrier-based airborne-early-warning aircraft or the C-2A *Grey-hound* carrier-on-board-delivery (COD) aircraft. This syllabus is an amalgam of T-34C, T-2C, and T-44A training which will qualify the student for both multiengine flying (T-44A) and carrier landings (T-2C).

The best helicopter pilots in the world are trained at Pensacola in the navy's rotary-wing program. Helicopter students go through the same primary and intermediate training as the maritime students. Then they break off to pursue their special brand of naval aviation, proceeding to Helicopter Training Squadron 8 (HT-8), based at the naval air station at Whiting Field, about twenty miles northeast of Pensacola.

Students in the maritime pipeline take advanced instruction in the Beech T-44 Pegasus. (Beech Aircraft)

At HT-8 the student spends six weeks studying aerodynamics and engineering phenomena peculiar to rotary-wing flight. He also logs some thirty-nine hours of flight time in the jet-powered TH-57 Sea Ranger before moving on to advanced training at HT-18, also based at Whiting.

Here the student is introduced to an advanced version of the TH-57 and in a sixty-six-hour flight syllabus becomes proficient in basic instrument techniques, radio navigation, cross-country flying, rough-terrain landings, night flying, rescue work, and formation flying. Carrier qualification and helicopter tactics are the finishing touches of training.

At this point each student pilot who has successfully completed his respective course of instruction will receive his navy wings of gold at a ceremony that may include friends and relatives. Henceforth, the pilot will be judged by a new standard of excellence— and he is more than up to it.

NAVAL FLIGHT OFFICER TRAINING

An equally important member of the naval aviation team is the NFO. In the fast-moving world of aerial warfare, he may find himself in the rear seat of a fighter as a radar intercept officer (RIO); a tactical navigator (TN) in an attack aircraft; a navigator (NAV); or a tactical coordinator (TACCO) on an antisubmarine, early-warning, or reconnaissance aircraft.

Upon completion of the AOC school or aviation indoctrination, the prospective NFO reports to Training Squadron 10 (VT-10) at Sherman Field in Pensacola, Florida, for fifteen weeks of basic training. A rigorous academic course covering all the basic operational aspects of naval aviation prepares the student for introductory flights in both the T-2 and the T-34C. Following basic training, NFO students split into various pipelines leading to specific warfare specialties. The categories of training are:

RIO radar intercept officer (fighters)
TN   tactical navigation (attack and electronic warfare)
OJN overwater jet navigation (carrier-based antisubmarine warfare and reconnaissance)

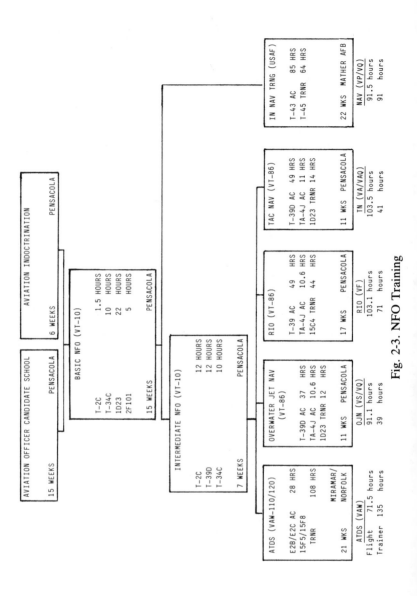

Fig. 2-3. NFO Training

ATDS   airborne tactical data system (early warning)
NAV   advanced navigation (patrol and land-based reconnaissance)

Students who are bound for patrol, reconnaissance, and other specialized land-based missions move on to Mather Air Force Base in Sacramento, California, for interservice navigation training in a course conducted jointly by the navy and air force. The course lasts twenty-two weeks and includes eighty-five hours in the air in the T-43, a training version of the Boeing 737 airliner.

The remaining students stay at VT-10 for seven weeks of intermediate training. Twelve hours in the T-2, ten hours in the T-34C, and twelve hours in the T-47 Citation quickly hone the all-important skills introduced in basic training. Instrument navigation on airways, instrument approaches, low-level/high-speed visual navigation, and crew coordination are all covered at this stage.

At the completion of intermediate training, those students destined for ATDS training and careers in the E-2 Hawkeye early-warning aircraft depart for either Norfolk, Virginia, or San Diego, California, to complete their undergraduate training in an E-2B or E-2C fleet replacement squadron. The remaining students move across the field to VT-86 for advanced training. Advanced training in VT-86 is given in three pipelines: RIO, TN, and OJN.

RIO training consists primarily of classroom, simulator, and airborne training to learn the basics of airborne intercepts and advanced airways navigation. This training includes forty-six flight hours in the T-47. Prior to designation as an NFO, all VT-86 students receive a short, intensive course in advanced tactical maneuvering. Several lectures and six flights in the TA-4J advanced jet trainer give the student a good feel for the problem.

The TN syllabus includes ground and flight training in low-level and radar navigation. Students spend about forty-nine hours in the T-47 and eleven in the TA-4J. The OJN syllabus grew out of the TN course to accommodate the specialized overwater training needs of NFOs going to S-3 Viking and A-3 Skywarrior squadrons.

Upon completion of the program, the student is designated an NFO and receives his wings of gold. He too will henceforth be judged by a new standard of excellence, for he, like the navy pilot,

A student NFO checks a radar image against his chart during a navigation problem in a flight simulator. (U.S. Navy)

will bear a large share of the responsibility for the successful prosecution of an assigned mission.

FLIGHT SURGEON TRAINING

Aviation medicine is a highly specialized field concerned with the adaptation of the human body to flight. As aircraft become faster and more maneuverable and reach greater heights, the practice of such medicine grows increasingly challenging.

Navy medical officers have the opportunity to become aviation medical specialists on the aeronautical team. The Naval Aerospace Medical Institute at Pensacola, Florida, provides a six-month course of instruction leading to the designation of flight surgeon. The institute trains up to 120 flight surgeons each year.

All flight surgeon trainees undergo three weeks of military orientation and indoctrination at the Naval Aviation Schools Command, fourteen weeks of classroom and clinical studies, one week of land and sea survival training, and six weeks of flight training.

Student flight surgeons receive special training and perform practical work in the medical fields relevant to aviation. They also learn to deal with problems such as stress, a common byproduct of flying.

The Naval Aerospace Medical Institute gives training in aerospace medicine, which is a specialty primarily concerned with the adaptation of man to flight. The trainee also learns the importance of the aerospace medical consultation service that supports the fleet. The successful performance of man in the abnormal environment aloft depends on the physiological and mechanical aids that the flight surgeon who becomes well indoctrinated in their use at the institute, is equipped with.

Vast programs established within the services and private industry to instruct aerospace medical specialists promise that careers in aerospace medicine will be amply rewarding, whether the physician is oriented toward research, military, clinical or preventive medicine.

The student flight surgeon course is divided into three phases. The first three weeks are devoted to military orientation and indoctrination at the Naval Aviation Schools Command. The student later returns to the command for one week of training in land and sea survival. The second phase of sixteen weeks is devoted to didactic and clinical studies, including special training and practical work in the fields particularly important to aerospace medicine, such as ophthalmology, otolaryngology, cardiology, neuropsychiatry, and physiology. The course also includes training and experience in the special problem areas created by the stresses of being an aviator or astronaut; trainees use such facilities as low-pressure chambers, ejection-seat devices, and the human centrifuge. The third phase of six weeks is devoted to flight indoctrination, during which the student flight surgeon receives limited hands-on flight orientation. As a "special crewman" in the fleet, he will participate in a wide variety of operational flying, including carrier landings and takeoffs as well as flights with land- and sea-patrol units and helicopter squadrons. Close association with naval and marine aviators, who are among the most select of men in uniform, is a rewarding experience in itself.

The flight surgeon wings, awarded upon graduation from this

course, signify that the physician is qualified to assume an important place in the aeronautical organization of the navy.

Naval aviation medicine is discussed in detail in chapter 10.

AIRCREWMAN TRAINING

The enlisted aircrewman is a proven performer and a highly skilled professional in his naval aviation specialty. Like the pilot, NFO, and flight surgeon, he flies by choice and is an important member of the flight team. On his skill may depend the detection of a hostile submarine, an incoming strike of enemy aircraft, or the location of enemy weapons and equipment to be neutralized.

The aircrewman receives formal orientation and indoctrination at the naval aircrew candidate school at Pensacola. This is the first of several training courses that will eventually lead to the designation of aircrewman. After a month in this intensive course, which includes physical fitness, aviation physiology, and water survival training, the candidate moves on to a fleet replacement squadron for hands-on instruction and ultimate assignment to a fleet squadron. For further information on aircrewmen, see chapter 11.

3

The Squadron

By Captain Richard C. Knott, USN

The squadron is the basic operational unit of naval aviation. It is a closely knit organization and less formal in some respects than other U.S. Navy commands. The squadron experience is characterized by a unique brand of camaraderie and a strong sense of esprit de corps, which promotes teamwork and significantly influences the quality and effectiveness of the squadron's performance. It is in this environment that the first-tour pilot or NFO gets his initial taste of operational flying.

Squadrons vary greatly in terms of equipment and the nature of their operations. Aircraft range from one- or two-place fighters to multiengine patrol planes with large, technically sophisticated crews. Some squadrons are carrier-based, spending part of their time at sea and part ashore. Others are land-based but deploy for several months at a time to overseas areas. Some, divided into detachments, are scattered to the winds. Still others, such as training squadrons, do not deploy at all. Whatever the case, each squad-

The Squadron (McDonnell Douglas)

ron type fulfills a necessary function, and all are essential to the successful execution of the navy's mission.

There are over two hundred active navy and more than fifty naval air reserve squadrons in existence today. The exact number changes constantly as new squadrons are established and old squadrons are disestablished to meet the changing needs of the navy. It should be emphasized here that the proper terminology for the origination or termination of a U.S. Navy squadron is establishment and disestablishment, respectively. The terms commissioned and decommissioned, used incorrectly in this regard, are properly applied to ships.

SQUADRON DESIGNATIONS

Squadrons are most often identified by letter-number designations that begin with the letter *V* or *H*. The prefix *V* has been with us since the early 1920s, when it was used to distinguish heavier-than-air from lighter-than-air aircraft, the latter type having been identified by the letter *Z*. The U.S. Navy has had no lighter-than-air aircraft since 1962, so this prefix is no longer used.

The squadron is a closely knit organization of men and aircraft. These Grumman A-6 Intruders are part of an attack (VA) squadron which, in turn, is part of a carrier air wing. (U.S. Navy)

The prefix *V* is still used to identify squadrons of fixed-wing, heavier-than-air aircraft. Squadrons of rotary-wing aircraft are identified by the letter *H* for helicopters.

The letter or letters that make up the suffix are used to signify the squadron's mission. For example, the suffix *T* stands for training. When combined, the letters *VT* identify a training squadron of fixed-wing aircraft. The letters *HT*, on the other hand, identify a training squadron of helicopters.

Two-letter suffixes may also be used. For example, the letters *VFA* make up the letter designation of a strike fighter squadron that has both an attack and fighter capability. Using another example, a light helicopter antisubmarine squadron is identified by the letters *HSL* (helicopter, antisubmarine, light).

To identify a specific squadron of a given type, numbers are added to the letters. Hence, VF-1 is the letter-number designation of Fighter Squadron 1.

Squadrons can also be identified by acronyms. Using this system, Patrol Squadron 30 (VP-30) becomes PATRON 30, while Air Test and Evaluation Squadron 1 (VX-1) becomes AIRTEVRON 1.

SQUADRONS AND THEIR MISSIONS

For purposes of discussion, the squadron types have been arranged below in four major categories—carrier squadrons, land-based antisubmarine squadrons, special mission/support squadrons, and training squadrons.

Carrier Squadrons

Each aircraft carrier has its own air wing. Each air wing in turn is made up of fighter, attack, antisubmarine warfare, tactical electronic warfare, and airborne-early-warning squadrons as well as special mission squadrons and detachments. Working together under the air wing commander, they provide the offensive striking power of the carrier as well as its primary means of defense.

Land-Based Antisubmarine Squadrons

Antisubmarine patrol squadrons are based ashore and can be deployed on short notice to virtually any area of the world. Although their primary mission is antisubmarine warfare, these squadrons have additional capabilities in such areas as ocean surveillance and minelaying. The addition of Harpoon missiles even provides an air-to-surface attack capability.

For administrative and training purposes these squadrons are assigned to patrol wings. When deployed, patrol squadrons operate under the appropriate operational commander in the area to which they are assigned.

Special Mission/Support Squadrons

Special mission/support squadrons provide a variety of services, including electronic reconnaissance, logistics support, utility services, helicopter services, electronic countermeasures, mine countermeasures, aircraft ferry services, air antisubmarine-warfare capabilities for smaller surface ships, and air test and evaluation.

Squadrons in this category are generally under the control of the functional wing commanders.

Table 3-1. Squadrons and Their Missions

Squadron/Aircraft Type	Letter Designation	Acronym	Squadron Mission
Attack/A-4	VA	ATKRON	Light attack; offensive and defensive air-to-surface attack operations with conventional and nuclear weapons.
A-6, A-7			Medium attack; all-weather offensive and defensive air-to-surface attack operations with conventional and nuclear weapons.
Fighter/F-4, F-5, F-14	VF	FITRON	All-weather offensive and defensive air-to-air operations to establish and maintain local air superiority and provide for fleet air defense.

Table 3-1. Squadrons and Their Missions—Continued

Squadron/Aircraft Type	Letter Designation	Acronym	Squadron Mission
Strike fighter/F/A-18	VFA	STRIKFITRON	Offensive and defensive air-to-surface attack operations with conventional and nuclear weapons; all-weather offensive and defensive air-to-air operations to establish and maintain local air superiority in the vicinity of a strike group.
Tactical electronic warfare/EA-3, EA-6, EA-7	VAQ	TACELRON	Tactically exploits, suppresses, degrades, and deceives enemy electromagnetic defensive and offensive systems, including communication, in support of air strike and fleet operations; aerial refueling services in some squadrons.
Carrier airborne early warning/E-2	VAW	CARAEWRON	Early-warning services for fleet forces and/or shore warning nets under all-weather conditions; command and control facilities for carrier air wings.
Light photographic reconnaissance/RF-8	VFP*	LIGHTPHO-TORON	Provides aerial photographic intelligence services for fleet operations.
Carrier antisubmarine warfare/S-3	VS	AIRANTISUB-RON	Carrier-based all-weather antisubmarine warfare.
Helicopter antisubmarine	HS	HELANTISUB-RON	All-weather antisubmarine operations.

Table 3-1. Squadrons and Their Missions—Continued

Squadron/Aircraft Type	Letter Designation	Acronym	Squadron Mission
Patrol/P-3	VP	PATRON	Land-based all-weather antisubmarine, surveillance, mine warfare, and maritime air-to-surface attack operations.
Fleet air reconnaissance/EA-3, EP-3	VQ	FAIRECONRON	Electronic warfare support including search for and interception, recording and analysis of radiated electromagnetic energy, in support of military operations. Selected squadrons serve as elements of the Worldwide Airborne Command Post system and provide communications relay services.
Fleet composite/various	VC	FLECOMPRON	Utility services. Air services for fleet training. Maintains combat capability in order to augment combat forces as necessary.
Aircraft logistics support/C-9, C-12, CT-39, C-130, C-1, C-2, US-3	VR/VRC	FLELOGSUP-PRON	Fleet logistics airlift. VR squadrons provide heavy airlift services. VRC squadrons specialize in carrier onboard delivery (COD).
Aircraft ferry/various	VRF	AIRFERRON	Aircraft ferry services in support of aviation activities.
Tactical aerial refueling/KA-3, KA-6	VAK*	AERREFRON	Day/night aerial refueling services in the Reserve forces.

Table 3-1. Squadrons and Their Missions—Continued

Squadron/Aircraft Type	Letter Designation	Acronym	Squadron Mission
Air test and evaluation/various	VX	AIRTEVRON	Tests and evaluates the operational capabilities of new aircraft and equipment in an operational environment. Develops tactics and doctrines for their most effective use.
Antarctic development/C-130, UH-1	VXE	ANTARCTIC-DEVRON	Supports operation DEEP FREEZE and in this capacity is under the operational control of the U.S. Naval Support Force, Antarctica.
Oceanographic development/RP-3	VXN	OCEANDEV-RON	Special oceanographic projects including geomagnetic surveys, Arctic Basin and Marginal Sea Zone surveys and oceanographic surveys in support of antisubmarine warfare and environmental prediction.
Helicopter attack squadron light/HH-1	HAL*	HELATKRON LIGHT	Helicopter gunship squadron. Provides fast-reacting tactical air support to Navy Special Warfare Groups and other Navy units.
Helicopter antisubmarine/SH-3	HS	HELANTISUB-RON	Conducts all-weather antisubmarine operations.
Helicopter anti-submarine light/ SH-3, SH-60	HSL	HELANTISUB-RON LIGHT	Provides LAMPS helicopter detachments aboard guided missile cruiser, destroyer, and frigate class ships. All

Table 3-1. Squadrons and Their Missions—Continued

Squadron/Aircraft Type	Letter Designation	Acronym	Squadron Mission
			weather antisubmarine warfare and antiship surveillance as well as targeting operations.
Helicopter combat support/UH-1, SH-3, UH-3, CH-46, UH-46	HC	HELSUPPRON	Helicopter search and rescue utility, logistic transport of nuclear weapons and vertical replenishment services in support of fleet requirements.
Helicopter mine countermeasures/CH-53, RH-53, MH-53	HM	HELMINERON	Provides detachments of helicopters specially configured for mine hunting, mine destruction and mine sweeping operations.
Training (fixed wing)/T-34, T-2, TA-4, T-45†, T-44, T-39, T-47	VT	TRARON	Basic and advanced training of student naval aviators and flight officers in fixed wing aircraft.
Helicopter training/TH-57, UH-1	HT	HELTRARON	Basic and advanced training of student naval aviators in rotary wing aircraft.

*Reserve squadrons only.
†To be introduced in late 1980s.

Training Squadrons

Training squadrons provide training for student pilots and NFOs. They are land based and fall under the administrative control of the chief of naval air training.

SQUADRON ORGANIZATION

Squadrons are organized into departments and divisions that are administered and supervised by the CO through the executive officer.

Patrol squadrons are normally made up of nine P-3 Orion aircraft and twelve qualified crews. Aircraft of Patrol Squadron 49 (VP-49) are shown lined up on the parking apron during a deployment to Keflavik, Iceland. (U.S. Navy)

The CO of a squadron may be either a pilot or an NFO. He is responsible for the efficient operation of his command and the successful execution of the squadron's mission. Among his responsibilities outlined in U.S. Navy Regulations are morale, safety, discipline, readiness, efficiency, and operational and employment orders.

A squadron CO is generally the rank of commander, although some large specialized squadrons may be commanded by captains. Whatever the case, he flies operational and training missions and maintains his qualifications along with other squadron officers. Indeed, as the most senior and experienced pilot or NFO, he is expected to lead the way in all major squadron events.

The executive officer is also a senior pilot or NFO, usually of the rank of commander. He is second in command and assumes the duties of CO when that officer is not present. He is responsible for administering the squadron and ensuring that it functions smoothly and efficiently.

Regardless of mission, all U.S. Navy squadrons have at least four departments—administrative, operations, maintenance, and

Grumman F-14 Tomcat aircraft of Fighter Squadron 14 (VF-14) in formation. (U.S. Navy)

safety. Beyond that, individual squadrons may have additional departments to suit specialized needs. Departments are divided into divisions and divisions into branches.

The administrative officer heads the administrative department and is responsible for such things as maintenance of officer and enlisted personnel records, control of classified material, and educational and legal services. He also supervises the activities of the first lieutenant who is, in turn, responsible for security, transportation, and the maintenance and cleanliness of squadron spaces and equipment.

The operations officer is charged with the proper planning and execution of squadron operations. His responsibilities encompass the areas of tactics, navigation, communications, and intelligence. He also supervises flight scheduling, training, the maintenance of operational logs and records, operational employment of nuclear weapons, and defense against nuclear, biological, and chemical warfare. In a carrier squadron, the LSO falls under his cognizance.

The maintenance department is typically the largest in the squadron. It is responsible, under the direction of a maintenance

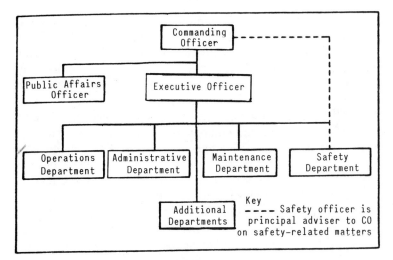

Fig. 3-1. Typical Squadron Organization

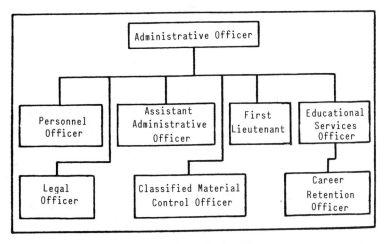

Fig. 3-2. Typical Administrative Department Organization

officer, for the upkeep of squadron aircraft and associated equipment at the organizational maintenance level. The technical complexity of today's weapons systems makes the maintenance officer's job an especially challenging one, as he must keep a maximum number of the squadron's aircraft in an "up" status if the organization is to perform its mission properly. His responsibilities include inspection and servicing and maintenance of airframes, power plants, electronic equipment, instruments, electrical systems, and other aircraft equipment. He supervises maintenance, material control, and quality assurance, and ensures proper handling of a large number of aircraft records, log books, directives, and reports. Through his line division officer, he directs the activities of plane captains, trouble shooters, and other ground support personnel. When an aircraft is placed in an "up" status, there is the implicit assurance of the maintenance officer that it is safe to fly and ready to execute the squadron's mission.

Aircraft and their associated systems are unforgiving of human carelessness. Recognizing this fact, the navy assigns an individual with full department-head status and direct access to the CO to every squadron to deal with the problem of safety. The safety

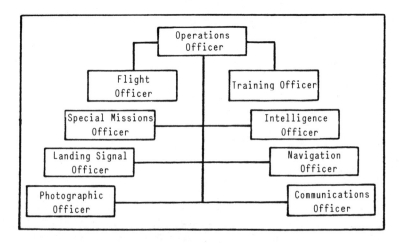

Fig. 3-3. Typical Operations Department

officer is typically on his second squadron tour and is a graduate of the navy's safety school at Monterey, California. He is a member of the squadron aircraft mishap board (AMB).

The most important part of the safety officer's job is the prevention of accidents. In this capacity he is most concerned with safety awareness. He prepares and implements an aggressive safety program and ensures that all personnel adhere to safety directives and employ safe working habits. NATOPS (naval air training and operating procedures standardization) publications prescribe safe standardized procedures for the operation of squadron aircraft and equipment. The safety officer may have one or more officer assistants to help him perform this vital squadron function.

FLEET-READINESS SQUADRONS

On completion of flight training and before proceeding to their first permanent squadron assignment, pilots and NFOs are ordered to fleet-readiness squadrons (FRSs) for training in a particular type of aircraft and mission. Enlisted aircrewmen and maintenance per-

Sikorsky CH-53E Super Stallions of Helicopter Combat Support Squadron 4 (HC-4) perform a variety of functions, including search and rescue, logistic transport, and vertical replenishment. (U.S. Navy)

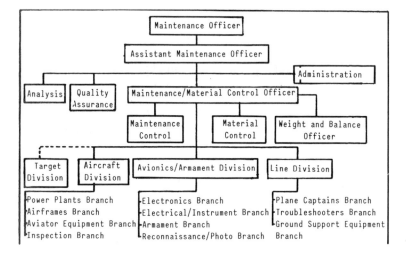

Fig. 3-4. Maintenance Department Organization

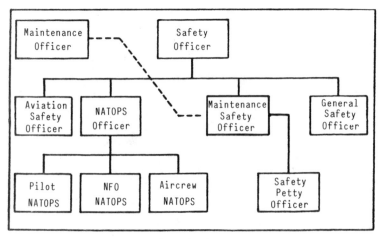

Fig. 3-5. Safety Department Organization

sonnel who have satisfied the requisite aircrew and technical course requirements will also be assigned to FRSs for aircrew and maintenance FRAMPS (fleet replacement aviation maintenance program) training. The length of the FRS courses for officers and enlisted personnel varies with the type of aircraft, the mission, and the function of the individual.

Table 3-2. FRS Training

Mission	Squadron	Type of Aircraft	Location	Course Length Pilots/NFOs
Attack	VA-42	A-6E	NAS Oceana	28/29 wks
	VA-128	A-6E	NAS Whidbey Island	28/29 wks
	VA-122	A-7E	NAS Lemoore	26 wks
	VA-174	A-7E	NAS Cecil	26 wks
Fighter	VF-101	F-14	NAS Oceana	30/30 wks
	VF-124	F-14	NAS Miramar	30/30 wks
Strike-fighter	VFA-106	F/A-18	NAS Cecil	19 wks
	VFA-125	F/A-18	NAS Lemoore	19 wks
Tactical electronic warfare	VAQ-33	A-3	NAS Key West	19/18 wks
	VAQ-129	EA-6B	NAS Whidbey Island	28/30 wks
Carrier-based airborne-early warning	VAW-120	E-2C	NAS Norfolk	24/33 wks
	VAW-110	E-2C	NAS Miramar	24/33 wks
Carrier-based antisubmarine warfare (fixed-wing)	VS-41	S-3A	NAS North Island	39/37 wks
	VS-40*	S-3A	NAS Cecil Field	39/37 wks
Patrol; land-based anti-submarine warfare	VP-30	P-3B	NAS Jacksonville	20/20 wks
	VP-31	P-3B	NAS Moffett Field	20/20 wks
Fleet air reconnaissance	USAF	EC-130Q	Little Rock AFB	8 wks
Fleet logistic support	----	UC-12B	NAS Norfolk	4 wks
	VRC-30	C-1 UC-12B	NAS North Island	4 wks

Table 3-2. FRS Training—Continued

Mission	Squadron	Type of Aircraft	Location	Course Length Pilots/NFOs
Helicopter antisubmarine warfare	HS-1	SH-3D SH-3G SH-3H	NAS Jacksonville	20 wks
	HS-10	SH-3D SH-3G SH-3H	NAS North Island	20 wks
Helicopter antisubmarine light	HSL-30	SH-2F HH-2D	NAS Norfolk	20 wks
	HSL-31	SH-2F HH-2D	NAS North Island	20 wks
	HSL-40†	SH-60B (LAMPS MK 3)	NAS Mayport	30 wks
	HSL-41	SH-60B (LAMPS MK 3)	NAS North Island	30 wks
Helicopter mine countermeasures	HM-12	RH-53D CH-53E MH-53E	NAS Norfolk	18 wks
Helicopter combat support	HC-16‡	UH-1N	NAS Pensacola	10 wks
	HC-3	CH-46E HH-46D	NAS North Island	16 wks

*Scheduled to be established in January 1987.
†Scheduled to be established 1986.
‡This squadron serves as search-and-rescue model manager.

Aircraft belonging to specific squadrons or units can be identified by one-letter, two-letter, or number-letter combinations displayed on vertical tail members. These codes are assigned by the chief of naval operations (CNO) in accordance with the aircraft visual identification system and are promulgated annually in the Naval Aeronautical Organization publication (OPNAV notice C-5400). Codes, once assigned to a squadron or unit, are rarely changed, even if an organization is transferred from one fleet to

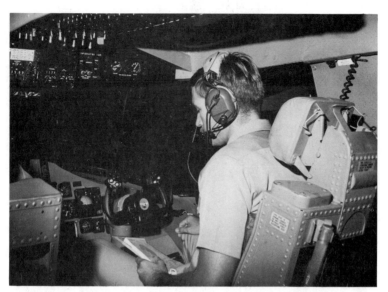

A replacement pilot gets a checkout in an operational flight trainer at FRS VAW-120. This simulator will introduce him to the Grumman E-2C Hawkeye before he actually flies that aircraft. (U.S. Navy)

another. When a unit is disestablished, its code becomes available for assignment.

Units belonging to AIRLANT (Air Forces, Atlantic Fleet) or AIRPAC (Air Forces, Pacific Fleet) have two-letter designations. The first letter of an AIRLANT unit code is taken from the first half of the alphabet (*A* through *M*), while AIRPAC units will take their first character from the second half (*N* through *Z*). The second letter in both cases may be any letter of the alphabet. The letters *I* and *O* are not used in either the first or second positions because they can be too easily mistaken for numerals.

Units under the chief of naval air training, including air stations, use single-letter codes taken from the letters *A* through *F*. Some air stations, naval air reserve units, units belonging to the chief of naval technical training, and overseas support units display number-letter combinations. Aircraft assigned to air stations or aircraft

An NFO at the FRS familiarizes himself with equipment he will be using during his operational tour. (U.S. Navy)

carriers (not squadrons) may also display the name of the unit, such as *Norfolk* or *Forrestal*.

Squadrons within a carrier wing can be readily identified by side numbers, usually displayed on the aircraft's nose and by the color often seen on some part of the tail fin.

Squadron	Side Number	Color
First VF Squadron	100–114	Insignia red
Second VF Squadron	200–214	Orange-yellow
First VA Squadron	300–315	Light blue
Second VA Squadron	400–415	International orange
Third VA Squadron	500–512	Light green
(strike) tanker	513–517	
VAW	600–603	Insignia blue
VAQ	604–607	Maroon
HS/HC (H-3)	610–617	Magenta
VS Squadron	700–713	Dark green
VFP	115–117	Black

(Source: OPNAVINST 3710.7).

THE BLUE ANGELS

One unusual navy squadron familiar to many Americans is the Blue Angels flight demonstration squadron. The Blue Angels trace their history back to 1946, when the team, made up of World War II veterans, began performing in Grumman F6F fighters. Later that same year they traded these aircraft for the highly maneuverable F8F, and in 1949 they made the transition to jets with the F9F Panther. A year later the team was organized into Fighter Squadron 191 and served in combat during the Korean War. In late 1951 it was reformed again, and in 1954 its base of operations was moved to Pensacola, Florida, where it remains today. The Blue Angels now fly the McDonnell Douglas A4 Skyhawk, whose performance and reliability have proven well suited to their task.

To be selected for assignment to the Blue Angels is a signal honor. There are only a few openings available each year, and competition is keen. Applicants must be career-oriented volunteers who have accumulated at least 1500 hours in tactical jet aircraft.

The Blue Angels Flight Demonstration Squadron in formation during a visit to the nation's capitol. (McDonnell Douglas)

The average Blue Angel is thirty-two years old (married or single), has at least four to six years of service in the navy or marine corps, and has completed at least one operational tour aboard a carrier.

While the Blue Angels are clearly an elite group, the basic flying techniques they employ are taught to every naval aviator. They have simply expanded their skills and honed them to a peak of perfection during one or more operational tours and a grueling team training schedule. Their performances in air shows across the country introduce the American taxpayer to naval aviation at its best and serve as an indispensable tool for recruiting the finest possible applicants.

4

The Aircraft Carrier

By Commodore Jeremy D. Taylor, USN

An aircraft carrier is an impressive sight by anyone's standards. A ship of gigantic proportions, it is a carefully engineered and highly mobile strike platform, an armored shell containing an extraordinary complex of powerful machinery and electronic sophistication. But a carrier is much more than a technological marvel. It is a floating city of some five to six thousand teammates who are trained to work and fight with split-second timing.

With its embarked air wing of seventy to one hundred tactical aircraft, the big-deck carrier is a weapons system of such awesome power that its very location or direction of movement on the world's oceans can have a significant effect on the course of international events. Mobile, flexible, and largely self-sufficient, it permits the United States to project its presence and power virtually anywhere in the world. With its tremendous war-fighting capabilities, the carrier is intended to deter aggression, and failing this, to prevail over the most powerful and determined adversary.

The aircraft carrier came into its own during World War II, and

The Aircraft Carrier (U.S. Navy)

by the end of 1946 the United States had more than one hundred carriers of all types in service. By that time the concept of war at sea had changed dramatically. It was clear that for the foreseeable future the aircraft carrier would be the focal point of naval warfare.

An important product of the war was the development of battle carriers (CVBs). Three of these, the *Midway, Franklin D. Roosevelt*, and *Coral Sea*, were commissioned after hostilities had ceased. The *Midway* and *Coral Sea* are still in service and will remain with the navy's battle groups into the 1990s, when the new nuclear-powered carriers *George Washington* and *Abraham Lincoln* are commissioned. Ships of the *Midway* class were larger and heavier than their predecessors, with armored flight decks and better compartmentalization for damage control. They were the transitional step between the *Essex*-class ships of World War II and the supercarrier of today. With modernization they continue as potent warships, their only disadvantage being that they carry fewer aircraft than the supercarriers.

After the war, the advent of jet aircraft, heavier and faster than airplanes, gave rise to the idea that the days of the aircraft carrier were numbered. This was not to be the case. Instead, existing carriers were adapted to deal with the new requirements. Flight decks were strengthened, more powerful steam catapults were installed, and jet-blast deflectors muted some of the new flight-deck hazards. Eventually the angle deck came into being, which permitted the carrier to launch and land high-performance aircraft simultaneously, and mirror and lens visual-landing systems replaced the LSO's "paddles."

The era of the supercarrier began with USS *Forrestal* (CVA 59). This ship, commissioned in 1955, was even larger than the *Midway*-class carriers; it had four catapults instead of two and four elevators instead of three. The *Forrestal* was followed by three sister ships, the *Saratoga, Ranger*, and *Independence*. The next class, the *Kitty Hawk*, only a slight variation on the *Forrestal*, includes the sister ships *Constellation, America*, and *John F. Kennedy*. Into the 1970s these attack carriers (CVAs) were backed up by about ten ships of the *Essex* and modified *Essex* class and the three *Midway*-class carriers; however, the role of the *Essex*-class carriers was changed to an antisubmarine mission, and they were designated CVSs from about 1950 until the early 1970s, when the carrier force was reduced to fifteen CVAs and then to twelve CVAs and CVNs.

The nuclear-powered supercarrier is clearly the most powerful and impressive of all carriers. The *Enterprise*, the world's first, was commissioned at Norfolk, Virginia, on 25 November 1961. The *Nimitz*, commissioned on 3 May 1975, and her sister ships, the *Dwight D. Eisenhower* and *Carl Vinson*, are all nuclear-powered carriers and improvements on the *Enterprise*. The nuclear-powered *Theodore Roosevelt* was christened in late 1984. Two other CVNs are under construction at Newport News, Virginia.

THE MODERN CARRIER

As of this writing, the United States has fifteen big-deck carriers, five of which are powered by nuclear reactors, and an *Essex*-class training carrier, the USS *Lexington*. The nuclear-powered carriers have the important advantage of being able to steam greater distances at high speeds without having to refuel. The nuclear-powered carrier is supplied with enough nuclear fuel to operate for

The aircraft carrier is an awesome weapon. Here Grumman F-14 Tomcat fighters fly over the USS *John F. Kennedy* (CV 67). (U.S. Navy)

more than twelve years without refueling. It is limited in its operations only by crew fatigue and the need to replenish supplies, jet fuel, and ordnance. Conventional carriers must have more than two million gallons of fossil fuel; the fuel on a nuclear carrier does not take up as much space, so there is more room for the crew, aircraft fuel, ammunition, and other consumables.

The modern, multipurpose, big-deck aircraft carrier embarks ninety to a hundred aircraft of differing types. These are the carrier's striking arm, and they have a long reach. Air-to-air refueling enables carrier aircraft to fly over 85 percent of the earth's surface. They can detect and locate hostile forces; attack and destroy surface, subsurface, and air targets at sea; support ground forces; and interdict enemy forces ashore when land-based air power is not available. Further, they can provide protective cover for land-based maritime patrol aircraft and other operations at sea.

The fourteen big-deck carriers afford flexibility in a variety of peacetime operations or limited-war situations as well as in the event of a major war, conventional or nuclear. They have a formidable nuclear-strike capability that an enemy must reckon with should the circumstances arise. A sixteenth carrier, the USS *Lexington* (AVT 16), whose home port is Pensacola, Florida, is used for training. In the event of a major war, she could be employed as a war fighter with the battle groups of the Atlantic Fleet. Whatever the circumstances, these ships are an essential element in the sea-power equation.

In an all-out conflict, the aircraft carrier would be the unit around which much of our naval might would be arrayed. Its purpose would be to ensure unimpeded use of the world's life-sustaining maritime arteries and to attack enemy objectives best reached from the sea. The modern aircraft carrier is equipped to deal effectively with the air, surface, and subsurface forces of any conceivable enemy. Further, because of its size, structural strength, extensive compartmentalization, massive protective armor, and sophisticated damage-control systems, it is the least vulnerable of all surface ships to sinking and destruction. It is also most likely to weather all but a direct hit or a very near miss.

Experience over the past thirty-five years suggests that limited

Crewmen ready an F-14 for launching aboard the USS *Forrestal* (CV 59). (U.S. Navy)

war is the type of conflict in which the United States is most likely to become involved. In 1950 the Communist forces of North Korea struck South Korea with lightning speed and captured most of the airfields capable of launching air strikes and supporting United

Nations troops on the ground. Carrier aircraft conducted all tactical air operations during the first stages of the Korean conflict and remained a critical part of the United Nations effort until the end of the war. The importance of attack carriers during this period was evidenced by an increase in their numbers from seven to nineteen. Carriers were also used to thwart mainland China's designs

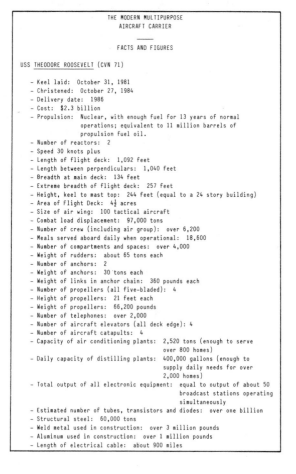

Fig. 4-1. The Modern Multipurpose Aircraft Carrier

on Formosa in the latter half of the 1950s. Carriers operating in the Gulf of Tonkin were indispensable to the conduct of the war in Vietnam, especially when the war was carried to North Vietnam during Operation Rolling Thunder and Line Backer.

It is generally considered that wars in the next few decades are likely to be those between smaller countries and third-world surrogates of a major power. In such conflicts the mobile, flexible, self-sufficient aircraft carrier reigns supreme in the projection of U.S. power. It has no peer.

It should be noted that the carrier need not actually engage in hostilities to produce the desired result. Aggressors who look upon neighboring countries as easy marks for a quick military takeover are likely to have second thoughts upon the appearance of a carrier battle group on the horizon. In this role, the job of the carrier is to prevent war. Steaming through narrow straits, cruising offshore, or lying at anchor in a foreign port, the aircraft carrier is a highly visible and impressive symbol of U.S. power—on call and ready on arrival anywhere, anytime. It is a comforting symbol to friends joined by forty-one separate treaties with the United States and a warning to potential adversaries. It says in no uncertain terms that the price of aggression may be unacceptably high.

ORGANIZATION

An aircraft carrier is organized much like other U.S. Navy combatant ships. The CO has absolute responsibility for the safety, well being, and efficiency of his ship. He is also provided with the authority to deal with his awesome responsibility The carrier CO is always a captain and is normally addressed as "The Captain" on a ship that may have as many as ten men in the rank of captain on board.

Carriers are all commanded by pilots or NFOs in compliance with the act of Congress that followed recommendations made by the president's aircraft board in November 1925. Chaired by Dwight Morrow, the board considered a wide number of aviation issues arising from the early military application of aircraft. The issue of carrier command was one of the most difficult questions considered by the board.

On the basis of testimony from military and civilian aviation leaders, the Morrow board recommended that selections for command or for general line duty on aircraft carriers and tenders, for command of flying schools, or for other important duties requiring immediate command of flying activities be confined to navy pilots. Congress enacted a law in 1926 (Public Law, Title 10, U.S. Code, Section 5942) that stated that the CO of an aircraft carrier or tender must be a line officer who is designated a naval aviator or naval aviation observer. A February 1970 amendment substituted the phrase naval flight officer for naval aviation observer.

The executive officer of an aircraft carrier acts as the CO's direct representative. He is primarily responsible for the organization, performance, good order, and discipline of the command and is prepared to assume command of the ship should the need arise. For this reason he would normally be stationed, during hostile action, in some position where he would not be subject to any casualty that might disable the CO.

Few billets in the navy are as demanding and difficult as that of executive officer of a carrier, or as effective in preparing a successful aviation squadron CO for ship and carrier command. A relentless schedule of operational and administrative activity must be carefully executed so that the operational readiness, safety, and

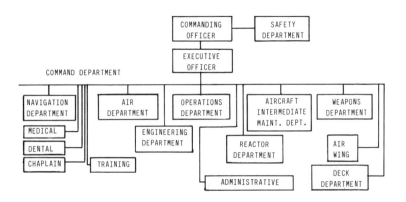

Fig. 4-2. Carrier Organization

The air boss is the focal point of all flight operations. Here he is seen in primary flight control aboard the USS *Enterprise* (CVN 65). (U.S. Navy)

welfare of the crew and material condition of the ship reach and remain at high levels. The executive officer is an experienced carrier airman who can effectively manage extensive human and material resources; he is also an expert in solving problems and getting things done.

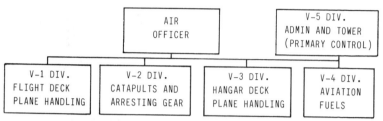

Fig. 4-3. Air Department

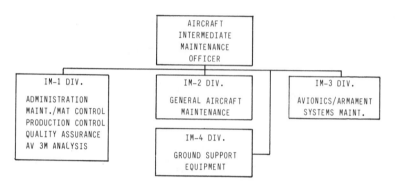

Fig. 4-4. Aircraft Intermediate Maintenance Department

The aircraft carrier is organized into departments and divisions for maximum efficiency. Unlike other navy ships, the carrier has an air department and an aircraft intermediate maintenance department. Further, the embarked air wing functions as a department, with the air wing commander as department head.

The air department is headed by the air officer, who is known in carrier parlance as the air boss. He is responsible under the CO for all air operations, including launching and recovery, and for the handling and servicing of aircraft aboard ship. His principal assistants are the assistant air officer or "mini boss," flight-deck officer, catapult officer, arresting-gear officer, hangar-deck officer, aviation-fuels officer, administrative assistant, aircraft-handling officer, and training assistant. The air department itself has five divisions which carry out department functions.

The aircraft intermediate maintenance officer, who heads a department of the same name, is responsible for the supervision and direction of the ship's maintenance effort in direct support of the embarked air-wing aircraft readiness. This organization is divided into one administrative and three production divisions, which are manned by skilled technicians whose primary function is to troubleshoot and repair state-of-the-art avionics, airframes, and jet engines used by squadron aircraft. Additionally, this department maintains and repairs ground-support equipment (yellow gear) used to maintain, start, and move embarked aircraft.

THE AIR WING

Each air wing (CVW) is typically made up of seven or nine squadrons of differing types of aircraft. This mix of aircraft provides the carrier's search, surveillance, and strike capability as well as fighter protection for the strike aircraft and the battle group itself. Included in the capabilities of the modern air wing are airborne early warning, electronic warfare, in-flight refueling, photo reconnaissance, antisubmarine warfare, and search and rescue. There are currently thirteen carrier air wings in active service, with two reserve carrier air wings available for emergency deployment. Additional wings will be added as new carriers are commissioned.

Table 4–1. Typical Carrier Air Wing

Number of Squadron	Function	Squadron Designation	Type of Aircraft	Total Number of Aircraft
2	Fighter and reconnaissance	VF	F-14 or FA-18	20–27
2	Light attack	VA	A-7 or FA-18	20–24
1	Medium attack (and tankers)	VA	A-6 and KA-6	10 plus 4
1	Electronic warfare	VAQ	EA-6	4 plus 4
1	Airborne early warning	VAW	E-2C	4 plus 4
1	Antisubmarine warfare	VS	S-3A	10 plus 4
1	Helicopter (ASW and SAR)	HS	SH-3H	6 plus 4

The wing is headed by the air-wing commander, who is called the "CAG." This acronym is a carryover from the early 1960s, when wings were called groups and the air-wing commander was known as the air group commander.

The air-wing commander is responsible for the leadership and management of the squadrons under his command. He supervises training and indoctrination, coordinates the activities of the squadrons, ensures material readiness, and oversees communications, intelligence, and other wing functions.

The air-wing commander is a unique warrior. As the oldest and most senior combat leader in the wing, he is first to meet the enemy. He thrives on tough missions, including night and all-weather flying. He leads. He is a tactician and operator, with a bold and aggressive fighting spirit tempered by judgment. He is savvy in naval warfare, including antiair warfare, antisubmarine warfare, and power projection. He is a weapons expert who knows how to fight and win. He is an experienced survivor in a profession unforgiving of human error. As an aviator who flies as many as seven different types of aircraft, he leads and directs the efforts of two thousand officers and men and is on track for even greater responsibilities. The CAG billet is the goal of every carrier aviator, whether he is a pilot or flight officer.

FLIGHT OPERATIONS

Preparations for flight operations usually begin the night before with the distribution of the air plan. An outline of the coming day's events, it includes launch and recovery times and information about the mission, number of sorties, fuel and ordnance loads, and tactical frequencies.

Flight quarters are announced to all hands over the ship's 1MC announcing system and manned as prescribed by the watch, quarter, and station bill. Crew members not directly involved in flight operations are not permitted on the flight deck or in the catwalks. Those who are have specific, clearly defined functions. They are recognizable at a glance by the colored helmets and jerseys that denote their roles.

Pilots and aircrews, meanwhile, have received comprehensive

briefings and understand clearly the sequence of events that will take place. They are usually ordered to man their aircraft forty-five minutes prior to launching. Before word is given to start en-

PERSONNEL	HELMET	JERSEY/ FLOTATION VEST	SYMBOLS, FRONT AND BACK
Aircraft handling crew and chock men	Blue	Blue	Crew number
Aircraft handling officers and plane directors	Yellow	Yellow	Billet title – crew number
Arresting gear crew	Green	Green	A
Aviation fuels crew	Purple	Purple	F
Cargo handling personnel	White	Green	"SUPPLY"/"POSTAL" as appropriate
Catapult and arresting gear officers	Green	Yellow	Billet title
Catapult crew	Green	Green	C
Catapult safety observer (ICCS)	Green	(Note 4)	Billet title
Crash and salvage crews	Red	Red	Crash/Salvage
Elevator operators	White	Blue	E
Explosive ordnance disposal (EOD)	Red	Red	"EOD" in black
GSE troubleshooter	Green	Green	"GSE"
Helicopter LSE	Red	Green	H
Helicopter plane captain	Red	Brown	H
Hook runner	Green	Green	A
Landing signal officer	None	White	LSO
Leading petty officers:			
Line	Green	Brown	Squadron designator and "Line COP"
Maintenance	Green	Green	Squadron designator plus "Maint. COP"
Quality assurance	Brown	Green	Squadron designator and "QA"
Squadron plane inspector	Green	White	Black and white checkerboard pattern and squadron designator
LOX crew	White	White	LOX
Maintenance crews	Green	Green	Black stripe and squadron designator
Medical	White	White	Red cross
Messengers and telephone talkers	White	Blue	T
Ordnance	Red	Red	Black stripe and squadron designator/ships billet title
Photographers	Green	Green	P
Plane captains	Brown	Brown	Squadron designator
Safety	White	White	"SAFETY"
Supply VERTREP coordinator	White	Green	"SUPPLY COORDINATOR"
Tractor driver	Blue	Blue	Tractor
Transfer officer	White	White	"TRANSFER OFFICER'

Fig. 4-5. Flight-Deck Personnel

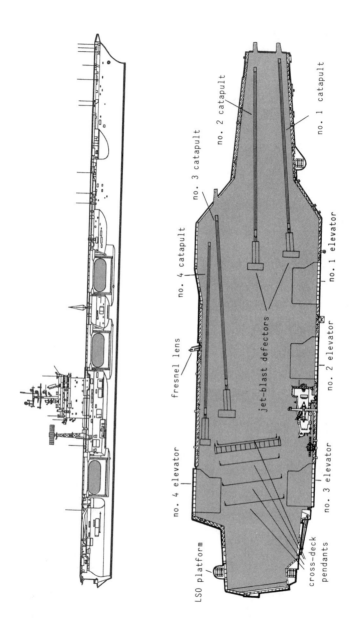

no. 2 catapult

no. 1 catapult

no. 3 catapult

no. 4 catapult

no. 1 elevator

fresnel lens

jet-blast defectors

no. 2 elevator

no. 4 elevator

no. 3 elevator

LSO platform

cross-deck
pendants

Fig. 4-6. The USS *Nimitz* (CVN 68)

gines they conduct preflight inspections of their aircraft to ensure that all is in order.

Flight-deck and squadron-maintenance personnel have also been busy readying their equipment and conducting a "FOD (foreign object damage) walkdown," in which they systematically comb every inch of the deck for loose material that could be blown about by jet engines or prop wash and cause injury to personnel or aircraft engines. Only after the FOD walkdown is the order given to start engines.

By this time the carrier has been turned to a course that will provide approximately 30 knots of wind over the flight deck for launch. Aircraft are directed forward by yellow-shirted plane directors and precisely positioned on the steam catapults. As an

A hook-up man signals the catapult officer that an F-14 is ready to launch. (U.S. Navy)

aircraft is "spotted" on the "cat," a blast deflector rises from the deck behind it to protect personnel and aircraft aft of the catapult.

A green-shirted hookup man attaches the nose gear of the aircraft to the catapult shuttle by means of nose-tow and holdback bars. When all is ready the pilot, on signal from the yellow shirt, releases his brakes and applies full power. At this time the catapult officer signals with a rotating hand motion, two fingers extended. After a final check to see if the aircraft is functioning correctly, the pilot salutes to indicate he is ready and braces himself for the shot. The catapult officer makes final checks on the cat's readiness and confirms from other on-deck personnel that the aircraft is ready for flight. He then touches the deck, signaling to a crewman in the catwalk to press the steam-catapult firing button. The aircraft is shot into the air, accelerating from zero to a normal "end speed" of 150 knots. The acceleration from zero to safe flying speed puts tremendous pressure on the aircraft and its crew. This spectacular achievement, which is usually duplicated more than one hundred times a day during peacetime carrier operations, is the product of brilliant engineering, careful and skilled maintenance of equipment, effective training of intelligent and motivated personnel, and attention to detail and safety.

An F-14 is launched from the no. 2 cat. (Grumman Aerospace)

As the aircraft becomes airborne, the catapult crews are already scrambling to position and hook up the next plane. A proficient team of four catapult crews can launch an aircraft every twenty to thirty seconds. In a matter of five minutes the ship can launch twenty aircraft and commence recovery operations.

The airborne aircraft, meanwhile, are under the control of the carrier-air-traffic control center (CATCC), which guides them in the carrier control area. As planes are launched, they join up at designated rendezvous areas and proceed to carry out various missions as directed by the air plan.

When the aircraft return to the ship and weather precludes a visual approach, CATCC controls their arrival and clears each

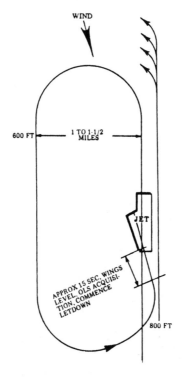

Fig. 4-7. Daylight Visual Landing Pattern (JETS)

aircraft for approach at one-minute intervals. It is the LSO who becomes the key player in assisting the pilot in his final approach to landing. The LSO is a carefully selected, seasoned carrier pilot—a "tailhooker"—who has had extensive training in this specialized field. He operates from a well-equipped platform abeam the landing area on the aft portside of the ship. He, his assistant, and the LSOs under training correlate factors such as wind, weather, aircraft characteristics, deck motion, and pilot experience to guide the pilot as his aircraft makes its final approach. The LSO is an expert, his judgment fine-tuned and rarely questioned.

For recovery in visual conditions, the aircraft return to an overhead "stack" at altitudes prescribed by air-wing doctrine. Individual flight leaders "take interval" on the flights at lower altitudes in the stack. Aircraft, in formations of two to four, normally enter the break for landing from astern of the ship, on the same heading and slightly to the starboard side, at an altitude of eight hundred feet. The flight leader will break left when he has reached a position projected ahead of the ship, establish himself in the downwind leg, descend to six hundred feet, and complete his landing checklist in preparation for landing. For his final approach, he will normally use the fresnel lens optical landing system (FLOLS), a combination of lenses and lights located on the port edge of the angled deck. This is an automatic, gyrostabilized system. If it should fail or if the ship is rolling and pitching beyond the limits of gyrostabilization, a manual optical visual landing aid system (MOVLAS) will be used instead. In good weather conditions, this entire operation—the recovery of approximately twenty aircraft—is conducted "zip lip," with no radio communication.

As the aircraft lines up for its final approach at somewhere between 120 and 150 knots, depending on the type of aircraft, the pilot will observe and fly the "meatball." The "meatball" is an amber light that appears at the center of a "mirror," which in reality is a stack of five lenses. If the aircraft is properly positioned on the glide path, the "meatball" will be aligned with a horizontal line of green reference lights on either side of the center lens. If it is above the glide path, the "ball" will appear on one of the upper lenses; if below, on one of the lower ones. The pilot's objective is to keep the ball centered all the way to touchdown and to engage the "three wire"—the third of four cross-deck pendants (wires) extending up the deck from the fantail, or the ramp, as "tailhookers" refer to the stern of the ship.

While it is the pilot's responsibility to fly the ball, the LSO may also give light signals or voice instructions until touchdown. If the approach is unsafe, the LSO will press his "pickle" switch, which activates flashing red lights and orders a wave-off. The pilot has no option in this situation; he must comply with the order to take his aircraft around the landing pattern again for another approach.

The aircraft will normally land on the angled deck, catch a wire, and be brought to a halt within a few hundred feet. The cross-deck wires are attached to cables that are reaved through pulleys and around the drums of the ship's four arresting-gear engines. Hydraulic dampeners are adjusted for each aircraft according to its weight, so that the arrestment does not exceed the aircraft's structural limits but does stop the aircraft within the landing area.

A right-seat view of a final approach as seen from the cockpit of a Grumman A-6 Intruder. (U.S. Navy)

As the landing aircraft makes contact with the deck, the pilot moves his throttle to the full-power position. If his aircraft's tail-hook engages one of the cross-deck wires, he immediately retards the power so the engines idle as the aircraft is brought to an abrupt stop. This is an arrested landing—a "trap." If the landing aircraft does not engage one of the four cables, the throttle remains in the full-power position, the engine accelerates, and the aircraft becomes airborne again for another try. This event, called a "bolter," is the principal reason for designing the carrier with an angled deck. If the aircraft has been successfully trapped, the plane is permitted to roll back a few feet so that the wire can be disengaged from the hook. The hook is raised and the aircraft is then taxied forward and parked on the bow. The arresting gear is quickly reset for the next aircraft. During recovery operations a proficient carrier air-wing team will complete the recovery of twenty aircraft in fifteen to eighteen minutes.

A pilot making a good approach snags either the number two or number three wire. If he catches the number four, he was probably high or fast on the glidescope; a trap on the number one

A Grumman E-2 Hawkeye lands aboard the USS *Kittyhawk* (CV 63). (U.S. Navy)

indicates that he was low or slow. The LSO grades every approach on a carrier trend analysis form so that there is a continuing record of each pilot's performance. Competition among pilots is keen.

When necessary, aircraft are recovered in bad weather. A variety of systems are available to aid foul-weather landings, including the instrument landing system, tactical air navigation system (TACAN), carrier controlled approach (CCA), and the automatic carrier landing system (ACLS).

The ACLS is capable of bringing an aircraft to touch down when a pilot has no visual contact with the landing area. A computer takes information from the ship's precision radar and sends signals to the aircraft's automatic pilot, which in turn flies the aircraft and executes the approach. In ACLS landings the pilot does not have to touch the controls.

THE PATH TO COMMAND

The path to carrier command, rather broad at junior levels, narrows considerably with seniority. Superior performance in repeated assignments as a carrier aviator or flight officer and as CO of at least one aviation unit, normally a carrier-based squadron, invariably precedes selection for major command of an amphibious or support force deep-draft ship. Selection for carrier command is made from the survivors of this winnowing process, and each year five to ten men are screened for such assignments. There is no more prestigious, responsible, or challenging captain's assignment in naval aviation. It is a goal that the most ambitious seek but only a few attain.

Contenders for carrier command will typically have on their record between eight hundred and twelve hundred carrier landings, between four thousand and six thousand flight hours, and six to eight extended cruises during three or four squadron and/or air-wing tours. Most will qualify as underway officer of the deck and command duty officer before completing a tour as a department head aboard a carrier. Some will also have served on a major afloat staff. Many will have served as air-wing commanders. Other carrier COs will ascend to that most highly sought position via selection and assignment as carrier executive officers.

From these billets, contenders for carrier command enter training pipelines that will prepare them for ship command. Before they take over a deep-draft major command, they will complete additional courses in shipboard engineering, ship handling, tactics, and leadership.

Some will assume command of amphibious force ships in the LPH, LPD, and LKA classes, while others will "skipper" support force ships such as those bearing the designations AOE, AOR, AO, and AFS. Operating with battle and amphibious readiness groups, they will make more than fifty port entries in these large, deep-draft ships (from 25 to 37 feet) and operate them in heavily trafficked waters, steaming both independently and in formation. All this will be done in numerous exercises that show the flag in several countries. In short, contenders for carrier command must master the heavy responsibility of a captain.

For those selected for nuclear-power training following completion of aviation squadron command, the route to carrier command is especially arduous. After thirty to thirty-six months of sea

There is no more prestigious, responsible, or challenging an assignment than commanding officer of an aircraft carrier. (U.S. Navy)

duty as squadron execs and then COs, they complete sixteen months of intensive nuclear-engineering training. Then they serve two to three years as operations and executive officers on nuclear-powered carriers. Carrier command ensures these extraordinary performers two or three additional years of sea duty.

Carrier command is likely to come to only those pilots or NFOs who set their sights early and who pursue, with unswerving perseverance and skill, the proven paths established by Congress in 1926.

FLAG SHIP

As the centerpiece of the battle group, every carrier is equipped to accommodate the battle-group commander and his staff. The battle-group commander is normally a rear admiral who commands either a carrier group or a cruiser-destroyer group. It is highly unusual for a carrier to be deployed away from home coastal waters without a battle-group commander and his staff embarked. The carrier thus becomes the host for the officers who plan, coordinate, and direct the operations of the entire battle group. These include strike, antiair, antisubmarine, and electronic warfare.

The embarked admiral and his staff, along with the carrier's crew and the air wing, constitute a team that must work in close harmony if the capability of the carrier is to be realized. This harmony is quickly developed during predeployment training— the four- to five-month period of underway and in-port training that precedes a carrier's forward deployment with either the Sixth or Seventh Fleet, which lasts six or seven months.

CARRIER ROTATION

After returning from a forward deployment, a carrier enters a period of restricted availability largely determined by its material condition. While this is normally a short period of two to four months, a carrier that has been repeatedly deployed for extended periods with minimum voyage repairs can require several additional months of shipyard repairs before commencing predeployment training.

New people and new equipment are integrated into the carrier during the first months after a cruise (deployment). This is the period of refresher training during which the crew drills for all contingencies and the air wing returns to the ship to regain carrier landing and operations proficiency. For the next three or four months, the tempo and complexity of training is intensified in preparation for a final test of ship and air-wing ability (the operational readiness evaluation, or ORE) to carry out the many missions assigned to a carrier.

The ORE is a stern test of all facets of the carrier's ability to conduct sustained combat operations. A simulated real-world scenario is provided by the evaluating staff—usually led by the embarked battle group commander—and the ship and air wing respond as if it were the real thing. Around-the-clock operations are sustained for several days. When the evaluation is complete, scores are assigned and the ship is certified for deployment.

The USS *John F. Kennedy* (CV 67) with Carrier Air Wing 1 embarked. Mobile, flexible, and largely self-sufficient, the aircraft carrier projects U.S. presence and power virtually anywhere on earth. (U.S. Navy)

An additional predeployment period of about one month is normally scheduled following the ORE to enable ship and air-wing personnel to top off provisions, wrap up personal and family affairs, peak material readiness, and solve other last-minute problems.

A carrier can be deployed for six or seven months and then undergo ten months to one year of turnaround maintenance and training, during most of which the ship and air wing must be ready to deploy within ninety-six hours. As the centerpieces of the navy's battle groups, the carriers are in high demand.

FIRST IN, LAST OUT

Aircraft carriers and carrier battle groups are the favorite instruments of presidents when they want to send a show of military presence and power to some area of the globe. Carriers are therefore the first ships in and last out in situations in which United States national interests are threatened. Carriers, which, unlike overseas bases, do not require host nation approval, can be "ready on arrival" to perform a variety of missions—presence, reconnaissance, show of force, or power projection. And because they can be easily resupplied, carriers are able to remain on station indefinitely.

The mighty aircraft carrier is the weapon of choice, lending support to diplomatic efforts to achieve peaceful solutions or favorable outcomes to problems involving our national interests.

5

The Naval Air Station

By Captain Paolo E. Coletta, USNR (Ret.)

EVOLUTION OF THE NAVAL AIR STATION

The first naval aviation "station" was in fact no more than an experimental camp established in August 1911 at Greenbury Point, Maryland, across the Severn River from the U.S. Naval Academy. Aircraft were housed in tent hangars, and the navy's first aviators were assigned there under Captain Washington Irving Chambers, who was then in charge of naval aviation. Operations had barely gotten under way when in January 1912 the camp was packed up and moved to North Island, San Diego, California, where the weather was more hospitable and where inventor Glenn H. Curtiss, who had built the navy's first aeroplanes, had established his winter base.

In May of that same year, the aviators and their equipment returned to Greenbury Point, where experiments in both day and night operations were continued. Other officers, including First Lieutenant A. A. Cunningham, who became the marine corps' first aviator, soon arrived for duty under instruction.

The Naval Air Station (U.S. Navy)

There were a number of accomplishments at the camp, including notable altitude, distance, and endurance flights. It was here too that the first fatal accident involving a naval aviator occurred when Ensign W. D. Billingsly, the navy's ninth aviator, was thrown from his aircraft by a sudden down draft. To prevent a recurrence of this unfortunate event, Glenn Curtiss designed a safety belt.

The first permanent facility was the naval aeronautic station established at Pensacola, Florida, in January 1914 for the purpose of conducting ground and flight training. The marine aviators at Greenbury Point went to exercise with the advance base unit off Culebra, while the aviation camp itself, including the navy contingent, was transferred to Pensacola to open a flight school.

When the United States entered World War I, naval air stations were established along the Atlantic Coast from Key West, Florida, to Halifax, Nova Scotia. In between, "rest stations" were created

The naval aeronautic station at Pensacola, established in January 1914, was the navy's first permanent air facility. (U.S. Navy)

which were, for the most part, merely beaches where seaplanes could come ashore to refuel. Some thirty stations were established on foreign soil in the Azores, Britain, France, and Italy, and in Central America a station was established to guard the Panama Canal.

With victory, naval air stations abroad were turned back to the host countries. The station at Pensacola, and another at San Diego that had come into being during the war, became permanent training bases. Other stations also continued in operation as the requirements of naval aviation grew.

Several naval reserve air stations were built in the 1930s, and in 1938 some sixty-eight million dollars was appropriated for a building program. Funds were also made available through the Works Progress Administration, other depression-era agencies, and the Civil Aeronautics Administration.

In September 1940 an agreement with Britain granted many

The naval air station at Hampton Roads was one of several air stations established during World War I. It survived postwar demobilization and is shown here with De Haviland DT-2 seaplanes of Torpedo Squadron 1 (VT-1) lined up along the seawall. (U.S. Navy)

overseas aviation base sites to the United States. Others were obtained from Pan American Airlines.

With the coming of war there rose an urgent need for more and better basing facilities for training, antisubmarine warfare operations, and support of carrier-based aircraft. Existing air stations were expanded and a score of new ones built throughout the country. Facilities for blimp operations were also increased. Overseas stations were built in Greenland, Iceland, Britain, and the Azores, and their emphasis was on convoy protection and antisubmarine warfare. Stations in North Africa supported the assault on Axis forces there and then in Sicily and Italy. Aircraft from bases in the Caribbean and in Central and South America flew endless hours of patrol against the submarine menace to shipping. Other stations sprang up on islands across the Pacific as the Allies drove toward the enemy's heart in the home islands of Japan.

Rapid decommissioning of many air stations followed the end of the war. However, by 1947 a naval air reserve program saved many of them from extinction and even called for some new construction. Meanwhile, jet aircraft demanded longer runways as well as improved fuel storage and maintenance facilities, which provided a new impetus for upgrading and growth. The Korean and Vietnam wars made it necessary to maintain and improve a respectable array of naval air stations and facilities at home and abroad, most of which remain in active service today.

MODERN NAVAL AIR STATIONS AND FACILITIES

Navy airfields support fleet users, aviation training activities, research and development centers, test centers, and the naval air reserve. There are currently some eighty-three active U.S. Navy airfields in use throughout the world, including thirty-four naval air stations and nine naval air facilities.

Naval air stations and facilities differ in size and capability as well as in the activities they support. Air stations offer the most extensive facilities and complete support, while naval air facilities are generally smaller in size and have less overall support capability. Stations and facilities are constantly changing to adapt to technological and organizational changes within the naval aviation

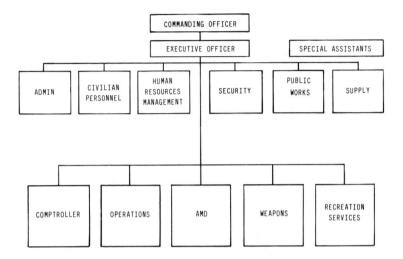

Fig. 5-1. Typical Organization of a Naval Air Station

community. Several are located on foreign soil, and some share runways with commercial airlines and the military services of host countries. Specific examples of various types of air stations and facilities have been selected for discussion in this chapter.

FLEET SUPPORT AIR STATIONS IN THE UNITED STATES

The air station at Moffett Field, California, is an example of one that has gone through several phases of development. Initially established as an air station in April 1933, it became home to the great dirigibles of that era, existed for seven years as an army air training base, served as a blimp center during World War II, and became a major naval air station again with the coming of jet aircraft.

After the demise of lighter-than-air activity at Moffett in 1947, the field became a helicopter overhaul and repair base and home to the navy's first atomic weapons squadron on the West Coast. It also served as the largest naval air transport base and the first of nine all-weather air stations. In 1950 it had the nation's first jet night-fighter squadron; later it became the first master jet base.

The naval air station at Moffett Field in the 1980s. Note the airship hangar built in the early 1930s. (U.S. Navy)

Because of noise problems, jet activities were ultimately transferred elsewhere. Moffett Field thereafter became host to P-3 Orion aircraft and headquarters for all land-based antisubmarine warfare operations in the Pacific. Today it provides support for the commander of Patrol Wings, Pacific (COMPATWINGSPAC), the commander of Patrol Wing 10 (COMPATWING 10), and a number of P-3 patrol squadrons, including VP-31, the FRS for the West Coast and Pacific area.

The air station at Cecil Field, Florida, is an East Coast air station that supports another kind of naval aviation activity. Located just west of the air station at Jacksonville, Florida, Cecil Field was an auxiliary airfield of that station in 1941 and became a naval air station in its own right in June 1952.

After World War II activity decreased and Cecil was almost closed, but in 1947 the navy began using the base for advanced fighter-pilot training. The following year it became an operating base for fleet aircraft units and received its first jets. Placed in

The naval air station at Cecil Field, seen from a Lockheed S-3 Viking aircraft on its final approach. (Hank Caruso)

partial maintenance status in 1950, it was saved by the Korean conflict and became one of the most active of the continental air stations.

Today Cecil Field is the home of the commander of Light Attack Wing 1 and a number of attack squadrons. It also hosts several S-3A antisubmarine squadrons.

Space limitations preclude discussion of the other fleet-support air stations. They are listed below, along with the major type of operational activity each supports.

Atlantic
    NAS Brunswick, Maine (VP)
    NAS Jacksonville, Florida (VP)
    NAS Key West, Florida (VF)
    NAS Norfolk, Virginia (HEL/VR)
    NAS Oceana, Virginia (VF/VA)

Pacific
    NAS Adak, Alaska
    NAS Alameda, California
    NAS Fallon, Nevada (VA)
    NAS Lemoore, California (VA/VFA)
    NAS Miramar, California (VF/VA)
    NAS North Island, California (HEL/VS)

Fig. 5-2. Naval Air Stations and Facilities in the Continental United States

NAS Whidbey Island, Washington (VA)
NAS Barbers Point, Hawaii (VP)
NAS Agana, Guam*

Naval air reserve stations and facilities are treated later in this chapter and in chapter 13.

FLEET SUPPORT AIR STATIONS OVERSEAS

The air station at Cubi Point in the Philippines, established in July 1956, is located within the confines of the U.S. naval base in Subic Bay. It has a 9,100-foot runway, hangars, and complete facilities for aircraft repair and servicing. It also has a 1,000-foot carrier pier that has hosted, at one time or another, every carrier now serving with the Seventh Fleet.

Cubi Point came into being largely as a result of the sponsorship of Admiral Arthur W. Radford, who urged its construction during the Korean War. Some three thousand Seabees leveled mountains and extended the land area into the sea to produce the desired

*Outside the United States but on U.S. territory.

A Lockheed P-3 Orion patrol aircraft is seen here over the naval air station at Barbers Point. (U.S. Navy)

result. It was a mammoth undertaking, but Radford's foresight has paid off handsomely in strategic benefits. During the Vietnam conflict, jets launched from Cubi were able to reach the Seventh Fleet's carrier striking force on Yankee Station in about one hour. Of even greater importance today is the fact that Cubi is within 1,500 miles of almost every critical point in the U.S. defense perimeter in the Far East.

The station at Sigonella, Sicily, located in the strategically advantageous center of the Mediterranean Sea, was originally established as a naval air facility in 1959. Because of its growing importance to Sixth Fleet operations, its support capabilities have been increased in recent years and its status upgraded to that of a naval air station in 1981. Its primary task is to provide aircraft maintenance and logistics services to the Sixth Fleet. To this end, a fleet logistics-support squadron and a helicopter combat-support squadron are based there. The Sigonella station is also an important base for land-based antisubmarine operations in the Mediterranean and hosts one deployed patrol squadron on a rotating basis.

Other overseas naval air stations are located on the island of Bermuda and at Guantanamo Bay, Cuba. The station in Bermuda is located about 825 nautical miles east of Savannah, Georgia, within range of most of the major North Atlantic sea-lanes. The former air force base is an important outpost for antisubmarine operations. P-3 Orion squadrons deployed there share the runways with commercial airliners whose terminal is located on the opposite side of the airfield.

Guantanamo Bay, Cuba, was one of the earliest sites used by the navy for operational deployment of the aeroplane outside the United States. Over the years it has played host to a variety of squadrons and detachments conducting exercises with the fleet.

Despite its long association with the aviation community, Guantanamo Bay did not become an air station until February 1941. It was used during World War II as a terminus for antisubmarine patrol seaplanes from the stations at Banana River (no longer in existence) and Jacksonville, both located in Florida. Station aircraft patrolled the approaches to Guantanamo Bay and conducted search and rescue operations as well.

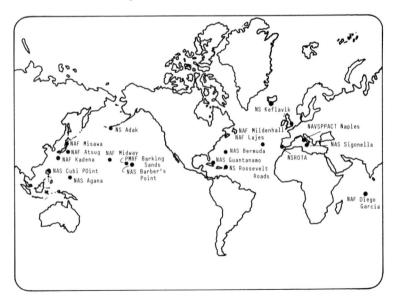

Fig. 5-3. Extracontinental Airfields

The Cuban Missile Crisis of 1962 saw an enormous airlift of troops and supplies to Guantanamo to repel a possible attack. Aircraft of all types used the facilities there in support of the quarantine operations imposed on the Soviets.

Today the primary mission of the Guantanamo Bay station is base and fleet logistics support and search and rescue. It also provides support for a fleet composite squadron and for numerous transient aircraft as well.

TRAINING AIR STATIONS

A formal training program for naval aviation personnel began with the establishment of the navy's first flying school at the naval aeronautic station at Pensacola, Florida, in January 1914. The station's first flight was made in a two-seat Curtiss biplane flying boat by Lieutenant John Towers and Ensign Godfrey Chevalier, the navy's third and seventh aviators, respectively.

Pensacola turned out aviators for World War I and was formally designated a naval air station in December 1917. During the twenties and thirties land planes used in carrier operations became increasingly important, and much of the training was carried out at Station Field (later Chevalier Field) and nearby Corry Field.

During the latter half of the thirties the Pensacola training complex expanded rapidly to accommodate the needs of the aviation cadet training program, established in 1935. This facility gave the United States a head start when it entered World War II. Even so, a number of auxiliary and outlying fields had to be constructed to absorb the increased flow of new aviators moving through the program.

After World War II several of the hastily added airfields were deactivated or placed in caretaker status, and the output of students was drastically reduced. When the Korean War broke out, Pensacola was again able to gear up quickly to accommodate the sudden demand for naval aviators.

Today the Pensacola station is the home of the chief of naval education and training (CNET). It is also host to the Naval Aviation Schools Command, the Aerospace Medical Institute, the Aerospace Medical Research Laboratory, the training carrier USS

Lexington (AVT 16), the Blue Angels, and the Naval Aviation Museum.

The headquarters of the chief of naval air training (CNATRA) is the air station at Corpus Christi, Texas. CNATRA is charged with the direction and coordination of the following training air wings:

NAS Chase Field, Texas	TRAWING 3
NAS Corpus Christi, Texas	TRAWING 4
NAS Kingsville,Texas	TRAWING 2
NAS Meridian, Mississippi	TRAWING 1
NAS Pensacola, Florida	TRAWING 6
NAS Whiting Field, Florida	TRAWING 5

These air wings are made up of some twenty squadrons, nine hundred aircraft, and fourteen thousand naval and civilian personnel.

The six naval air stations listed above are all primarily used to train officer personnel in flight duties. The station at Memphis, Tennessee, supports the Naval Air Technical Training Center (NATTC),* which is responsible for the instruction of enlisted personnel in aircraft maintenance and other aviation skills. Men and women who will be working in a variety of aviation fields begin with an aviation fundamentals course at Memphis and then move on to class A school training in individual rating specialties, also taught there. After one or more tours in a squadron or in some other fleet unit to gain practical experience, many petty officers return to the air station at Memphis for advanced training in their ratings. Maintenance officer courses are also taught there.

RESEARCH AND DEVELOPMENT AND TEST CENTERS

There are five research and development and test centers. Two of these, the Naval Air Test Center at Patuxent River, Maryland, and the Pacific Missile Test Center at Point Mugu, California, have naval air stations, the primary purpose of which is to support test activities.

For a number of years before World War II, the navy sent its

---

*The commander of the NATTC is headquartered at Lakehurst, New Jersey.

aircraft for tests of various kinds to air stations at Anacostia, Dahlgren, Norfolk, Philadelphia, and the Washington Navy Yard. As
aircraft and aircraft weapons systems became more complex, this
proved to be an unsatisfactory arrangement, especially in wartime,
and in January 1941 a flight test center was authorized. A site was
selected some sixty miles southeast of Washington, D.C., and the
naval air station at Patuxent River was built there in 1942–43.

The primary mission of the air station is to maintain and operate
facilities and to provide services and materials to support the Naval
Air Test Center. This includes support for the testing of strike,
antisubmarine, and rotary-wing aircraft as well as systems engineering, technical support, and computer services activities. The
U.S. Naval Test Pilot School is also part of the Naval Air Test
Center.

In a similar manner, the naval air station at Point Mugu, Cal-

Four test aircraft fly in diamond formation over the Naval Air Test Center,
Patuxent River. (U.S. Navy)

ifornia, supports the Pacific Missile Test Center, whose mission is the development, testing, and evaluation of naval weapons, especially missiles, and the provision and maintenance of the missile test range. The Pacific Missile Range Facility at Barking Sands, Hawaii, is a subordinate command of the missile test center and has its own airfield. It provides range support for the fleet and other U.S. government agencies that use the missile range.

Three other research and development centers, without naval air stations, have supporting airfields to serve their needs, the Naval Weapons Center at China Lake, California; the Naval Air Engineering Center at Lakehurst, New Jersey; and the Naval Air Development Center at Warminster, Pennsylvania.

NAVAL STATIONS AND NAVAL AIR FACILITIES

There are three naval stations with airfields and eleven naval air facilities throughout the world:

NAVSTA Keflavik, Iceland
NAVSTA Roosevelt Roads, Puerto Rico
NAVSTA Rota, Spain
NAF Atsugi, Japan
NAF Detroit, Michigan*
NAF Diego Garcia, Indian Ocean

Because of its proximity to the nation's capital, this naval air facility is sometimes referred to as the crossroads of the navy. (U.S. Navy)

NAF El Centro, California
NAF Kadena, Okinawa, Japan
NAF Lajes, Azores, Portugal
NAF Mayport, Florida
NAF Midway Island
NAF Mildenhall, England
NAF Misawa, Japan
NAF Washington, D.C.*
The naval support activity in Naples, Italy, shares an airfield with commercial aviation activities of the host country.

OUTLYING FIELDS AND AUXILIARY AIRFIELDS

Outlying fields and auxiliary landing fields are generally associated with specific naval air stations and are primarily used for training and overflow activities. These airfields are listed below with the naval air station indicated in parentheses.

Outlying Fields

OLF Alpha (NAS Meridian)
OLF Barin (NAS Whiting)
OLF Bravo (NAS Meridian)
OLF Brewton (NAS Whiting)
OLF Bronson (NAS Pensacola)
OLF Choctaw (NAS Pensacola)
OLF Coupeville (NAS Whidbey Island)
OLF Harold (NAS Whiting)
OLF Holley (NAS Whiting)
OLF Imperial Beach (NAS North Island)
OLF Middleton (NAS Whiting)
OLF Pace (NAS Whiting)
OLF Santa Rosa (NAS Whiting)
OLF Saufley Field (NAS Whiting)
OLF Silver Hill (NAS Whiting)
OLF Spencer (NAS Whiting)
OLF Summerdale (NAS Whiting)
OLF Whitehouse (NAS Cecil)
OLF Wolf (NAS Whiting)

*Naval air reserve facilities.

Auxiliary Landing Fields
   ALF Cabiness (NAS Corpus Christi)
   ALF Crows Landing (NAS Moffett)
   ALF Fentress (NAS Oceana)
   ALF Ford Island (NAS Barbers Point)
   ALF Goliad (NAS Chase Field)
   ALF Orange Grove (NAS Kingsville)
   ALF San Clemente (NAS North Island)
   ALF Waldron (NAS Corpus Christi)
Others
   Site 6 (NAS Pensacola)
   Site 8 (NAS Whiting)

NAVAL AIR RESERVE STATIONS AND FACILITIES

There are six naval air reserve stations and two naval air reserve facilities, the primary purpose of which is support of training for air reserve squadrons and aviation activities. Units supported include attack, fighter, patrol, logistics support, antisubmarine helicopter, and reconnaissance squadrons. These air stations and facilities are located in the following areas:
   NAS Atlanta, Georgia
   NAS Dallas, Texas
   NAS Glenview, Illinois
   NAS New Orleans, Louisiana
   NAS South Weymouth, Massachusetts
   NAS Willow Grove, Pennsylvania
   NAF Detroit, Michigan
   NAF Washington, D.C.

COMMAND OF A NAVAL AIR STATION

The modern naval air station is a major shore command. The CO has a challenging job that encompasses a wide range of responsibilities. He oversees the station's operations and ensures that it is providing effective support to tenant squadrons and other aviation activities. He is involved with all facets of air station affairs, from air operations to funding and construction. Outstanding management abilities are essential.

In carrying out his duties, he encounters many of the problems faced by a shipboard CO, particularly with regard to military personnel. Unlike the CO of a warship, however, he must also deal with a large number of civilian personnel on a day-to-day basis.

The location of a naval air station is permanent, and possibilities for misunderstanding and friction with area residents are ever present. It is one of the CO's tasks to see that relations between the air station and the civilian community remain harmonious. He must be sensitive to local problems and potential irritants and prepared to head off confrontations. Rush-hour traffic, the impact of military personnel dependents on local school systems, labor complaints, jet noise, and accidents are but a few of the complications an air station CO may be called upon to deal with. An effective CO will take a positive approach to community relations from the outset to foster cooperation on a day-to-day basis and will become personally involved in worthwhile local activities.

Experience gained during a command tour of a naval air station will help prepare a senior officer for even greater responsibility. Indeed, selection for naval air station command may open an alternative route to flag rank.

6

A Naval Aviation Career

Revised by Captain J. J. Coonan, USN

During their first few years of service, many officers decide to make long-term personal commitments to naval aviation. Promotion opportunities, early responsibility, advanced education and training, and a high degree of job satisfaction are among the more frequent reasons given by those who choose to make naval aviation a career. This chapter describes career specialties and discusses typical assignment patterns as well as professional development opportunities of interest to those who are considering a career in naval aviation.

NAVAL AVIATION DESIGNATORS

All officers in the naval aviation community have designators that begin with the number 13. The 13XX community is made up of pilots, designated either 1310 or 1315, depending on whether the officer is a regular or reserve, and NFOs, designated either 1320 or 1325. All these officers have some facet of naval aviation as a primary career pursuit. They make up approximately one half of the unrestricted-line officers of the navy. Additionally, there are a significant number of limited-duty officers and chief warrant officers who support the aviation effort. A summary of naval aviation designators follows:

131X—Line officer qualified for duty piloting naval aircraft
132X—Line officer qualified for duty as a weapons operator/ mission specialist in naval aircraft
137X—Line officer in training for designation as an NFO
139X—Line officer in training for designation as a naval pilot
151X—Aeronautical engineering duty officer (AEDO)
152X—Aeronautical maintenance duty officer (AMDO)
163X—Air intelligence officer (AIO)
181X—Oceanographer
630X—Pilot (flying limited duty officer, or FLDO)
631X—Aviation deck LDO
632X—Aviation operations LDO
633X—Aviation maintenance LDO
636X—Aviation ordnance LDO
638X—Aviation electronics LDO
639X—Air traffic control LDO
731X—Aviation boatswain (warrant officer, or CWO)
732X—Aviation operations technician (CWO)
734X—Aviation maintenance technician (CWO)
736X—Aviation ordnance technician (CWO)
738X—Aviation electronics technician (CWO)
739X—Air traffic control technician (CWO)

The fourth number of the designator means the following:

0—Regular navy officer, permanent grade ensign or above
1—Regular navy officer, permanent grade WO
2—Regular navy officer, with temporary commission whose permanent status is enlisted
3—Regular navy officer on the retired list
4—Reserve navy officer, whose permanent status is enlisted
5—Naval reserve officer (except specified naval reserve officers on active duty in the training and administration of reserves, or TAR, program) whose permanent status is WO or higher

6—An officer of the naval reserve who was appointed in the naval reserve integration program from enlisted status

7—Naval reserve officer on active duty in the TAR program; includes officers of the TAR program rotated to billets other than TAR billets

8—Naval reserve officer, whose permanent status is a WO

9—Naval reserve officer on the retired list

NAVAL AVIATOR (PILOT)

A naval aviator is defined in article 1410100 of the Naval Military Personnel Manual as a "commissioned line officer in the Navy or Marine Corps who has successfully completed the course prescribed by competent authority for naval aviators." Exactly what is that course of instruction, and what difference in training separates the naval aviator from the officer who earns his pilot wings in another service? The answer is the training and qualification as a navy carrier pilot. It is carrier training and experience that makes naval pilots unique and establishes a common bond between the World War II pilots who turned the tide at the Battle of Midway, the tactical-air pilots who dotted the skies over Hanoi and Haiphong during the Vietnam War, and today's pilots who are flying the FA-18 or the F-14.

NAVAL FLIGHT OFFICER

NFOs are unrestricted line officers qualified to operate the sophisticated airborne weapons systems in our modern navy aircraft. They are often referred to as the "new breed," even though they have been around almost as long as naval aviation itself. As far back as 1912, flying officers other than pilots were carried aloft to operate and test new equipment. The need for naval aviation officers, or NAOs, as they were designated, was not great, and as a result the NAO program sustained itself strictly through volunteers. During World War II many aviation candidates found themselves being trained to be bombardiers, navigators, and meteorologists. Their outstanding performance indicated a continued need in naval aviation for qualified officer aircrewmen. Postwar

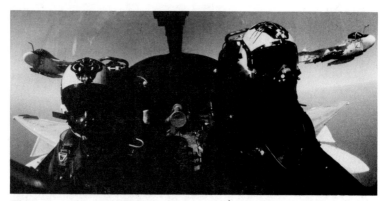

This navy pilot and NFO are a formidable attack team. Their career potential includes squadron, air wing, and carrier command as well as a variety of subspecialty opportunities. (U.S. Navy)

cutbacks in aircraft and personnel gradually sent the NAO program back into a state of dormancy. Not until the Korean War did the need for NAOs become evident again. After the hostilities ceased, the program did not decline, as it had before, but expanded with the introduction of new, sophisticated aircraft and technology.

In the mid-fifties technological advances in aircraft, ships, and submarines also necessitated a change in the tactical employment of forces, which placed a heavier burden on the pilot who was expected to fly his aircraft and simultaneously operate newly developed electronic sensors. As the state of the art progressed, it became readily apparent that an additional officer should manage the complex weapons systems.

During that period, the COs of squadrons employing heavy-attack, early-warning, and antisubmarine aircraft were hard pressed to find qualified personnel to fill the positions created by avionics advancement. Personnel planners soon realized that assigning pilots to these positions, which is what they were doing, was not the answer. While temporarily solving the manpower problem, it created morale problems among the aviators that would have been difficult to rectify had the policy continued. As an interim measure, the small core of naval aviation observers was expanded by a mixed group of nonpilot officers. Supplementing these officers were highly motivated first-class and chief radiomen and radarmen, some ex-pilots who were no longer qualified to fly because of their vision, and many "1100-designated" volunteers who expressed interest in the program and were physically qualified. There was little formal training in the program, which was sustained and enlarged through need alone. It became obvious that a specific program had to be established to procure and train candidates to fill the necessary positions. Finally, the NFO program came into being.

In July 1960 a basic naval aviation officers school was established at the naval air station in Pensacola. The school began with an eight-week training syllabus and sixty-five students as part of the training department of the naval air station. On 1 August 1963 it became a commissioned unit to train and indoctrinate nonpilot officers and officer candidates and to prepare them for advanced flight and technical training leading to the NFO designation. In August 1964 the curriculum was expanded to its present size and duration.

Two significant milestones in the development of the NFO program were passed in 1964 and 1965. First, it was determined that an NFO, like a pilot, would remain on permanent flight pay whether on sea or shore duty. Secondly, a separate designator, 132X, was assigned to take the NFO out of the general category of aviation officers who were not pilots. Subsequent congressional legislation removed the final barrier in the NFO career pattern by making NFOs eligible for command at sea, like their brother aviators. NFOs presently command one third of our aviation squadrons.

The NFO program is a popular choice for aviation candidates who want a challenging and rewarding career.

LIMITED-DUTY OFFICER

Limited-duty officers perform technical duties:
— Limited to specific occupational fields.
— That require authority and responsibility greater than that normally expected of a chief warrant officer.
— That require strong managerial skills.
— Outside a normal development pattern for unrestricted line

(URL) or restricted line (RL) officers (i.e., those duties that require excessive technical/on-the-job training by URL/RL officers).

The limited-duty officer meets the need for technical management skills. By filling these requirements, he frees URL officers for broader training. His officer management skills are an outgrowth of a technical background, and his duties are primarily in a supporting role and outside the normal development path of URL or RL officers. Such duties are defined as those for which a URL or RL officer would require extensive functional training and excessive on-the-job training and experience, or for which there are no URL or RL officers available.

Limited-duty officers, except for pilots (630X), are selected from the ranks of first-class petty officers and chief petty officers (in pay grades E6 to E8) with eight to sixteen years of active naval service. Pilots are selected from the ranks of second-class petty officers through chief petty officers. A small number of chief warrant officers who have served at least two years in their rank may apply for the limited-duty-officer program and, if selected, are commissioned in the grade of lieutenant junior grade.

The normal paths of advancement from enlisted status into the limited-duty-officer program are shown below:

Rating	CWO Category Designator	LDO Category
All		Aviator (630X)
ABE, ABF, ABH, AB	Aviation boatswain (731X)	Aviation deck (631X)
AW	Aviation operations tech (732X)	Aviation operations (632X)
AD, AME, AMH, AMS, AM, PR, AS, AZ, AFCM	Aviation maintenance tech (734X)	Aviation maintenance (633X)
AO, GMT	Aviation ordnance tech (736X)	Aviation ordnance (636X)
AC	Aviation traffic control tech (739X)	Aviation traffic (639X)

FLYING LIMITED-DUTY OFFICER (FLDO)

The flying limited-duty officer community (630X) was developed to help offset URL pilot shortages and provide stability in the aviation training command. Assignments are mainly in the primary flight training squadrons and ship's company billets aboard aircraft carriers. This career pattern was specifically developed to permit more URL pilots to fill operational billets instead of support functions. After selection, the selectee is required to attend AOC school and flight training. To qualify for the FLDO program, an enlisted member must be an E-5 through E-7; have more than four years of naval service; be under thirty; pass an AQTFAR with 3.5; have sixty semester hours of college or service equivalent; and pass a flight physical.

CHIEF WARRANT OFFICER

Chief warrant officers (designator 73XX) are technical officer specialists who perform duties limited in scope (in relation to other officer categories), technically oriented, and not significantly affected by advancement in rank (therefore amenable to successive tours). The chief warrant officer provides officer technical—as opposed to management—expertise at a relatively constant grade level in the naval officer hierarchy. Those who advance in rank within the chief-warrant-officer structure do not significantly widen their management function. Career development is focused on increasing the technical competence of the chief warrant officer within his specialty.

Chief warrant officers are selected from the ranks of our outstanding chief petty officers (pay grades E7 through E9) with twelve to twenty-four years of active naval service. All chief warrant officers in the U.S. Navy are commissioned officers in the grades of W-2, W-3, or W-4 (the noncommissioned or warranted grade W-1 is no longer in use).

Chief warrant officers are able to provide a valuable service to the navy. In support of the aviation navy they serve in six technical areas and bring the knowledge and experience gained in their enlisted ratings. Chief warrant officers in support of naval aviation are assigned the following designators:

Enlisted Rating	CWO Category
ABE, ABF, ABH, AB	Aviation boatswain (731X)
AW	Aviation operations tech (732X)
AD, AME, AMH, AMS, AM, PR, AS, AZ, AFCM	Aviation maintenance tech (734X)
AO, GMT	Aviation ordnance tech (736X)
AT, AX, AQ, AE, TD, AVCM	Aviation electronics tech (738X)
AC	Air traffice control tech (739X)

AERONAUTICAL ENGINEERING DUTY OFFICER

For those who have a technical bent, the aeronautical engineering duty officer (AEDO) program offers specialized career opportunities. This program is for aviation officers with nine to twelve years of proven performance, at least two tours of operational experience, and the requisite educational qualifications (usually a graduate degree). Application for the AEDO program is made through the Naval Military Personnel Command (NMPC-211) at least sixty days prior to a meeting of the Restricted Line Transfer Board, which convenes twice a year. Most billets for AEDOs are in aeronautical engineering or related fields. A number of command opportunities exist, primarily at the captain level, within the Naval Air Systems Command at naval air rework facilities, naval plant representative offices, and other field activities. There are about four hundred AEDOs on active duty in naval aviation.

AERONAUTICAL MAINTENANCE DUTY OFFICER

Aeronautical maintenance duty officers (AMDOs) come directly from the AOC program or are officers already serving in some other area of naval aviation. Those who come directly from AOC school split off from their pilot and NFO classmates immediately after commissioning and receive sixteen weeks of specialized training at the Aviation Maintenance Officers School at Pensacola. From there they are ordinarily assigned to sea duty in an organizational (squadron) level maintenance billet and thereafter will alternate on shore and sea duty tours through the rank of commander. The other source of AMDOs is proven aviation officers who enter by way of the Restricted Line Transfer Board in much the same way as AEDOs. A college degree is required for all AMDOs, regardless of source. As with the AEDO program, a number of command opportunities exist, primarily at the captain level, in various maintenance areas. There are about six hundred AMDOs on active duty in naval aviation.

AIR INTELLIGENCE OFFICER

Air intelligence is really a subspecialty in the field of intelligence. A junior officer entering this field through air intelligence training can expect his first two or three tours to be associated with naval aviation; he will normally start as a squadron air intelligence officer. Subsequent tours will broaden his professional knowledge in other interesting fields of naval and joint-service intelligence.

AVIATION TRAINING

All pilots and NFOs begin their naval aviation training in Pensacola, Florida. There, at the Naval Aviation Schools Command, they make decisions that will influence their entire navy career. Pilots must decide what general type of aircraft they want to fly, jet, prop, or helicopter. For NFOs, the aircraft type will be selected. The selection process is competitive. An aviator's entire performance, from the initial day of training, is important. Obtainment of a preferred aircraft type is directly affected by his competitive standing in the class. Only when a specific training pipeline has been selected (jet, prop, or helicopter training) do individuals continue to compete for specific aircraft assignment upon earning their wings of gold. The selection of aircraft type should be made carefully, since it will likely determine the aviation officer's whole naval career. His future assignments are frequently governed by the aircraft type he has chosen and the location of bases associated with that type. All fighters are based at either Oceana Naval Air Station, Virginia Beach, or Miramar Naval Air Station, San Diego; all East Coast patrol aircraft are based at either the naval air station in Brunswick, Maine, or the naval air station in Jacksonville, Florida; and all West Coast patrol aircraft are based at either the naval air station at Moffett Field or the naval air station at Barbers Point, Hawaii.

Although pilots and NFOs may serve together once they have reached an operational squadron, there are some major differences in their training progression. Student pilot training takes twelve to eighteen months to complete, depending on the type of aircraft, the student load in the training pipelines, and other unpredictable circumstances. As a general rule, the student pilot can count on slightly more than one year in various training squadrons. Upon completion of the training command syllabus, the coveted wings of gold are awarded. The junior pilot then proceeds to an FRS as a ready but as-yet unpolished "nugget." It is in the FRS that the training in operational fleet aircraft is given. Once this training is complete, the junior pilot is ready for assignment to a fleet squadron.

The training command syllabus for NFOs is slightly shorter than that for pilots because less flying time is required of NFOs. NFO training takes approximately one year to complete. Upon completion, wings are awarded and the newly designated officer is transferred to the FRS. Once in that squadron, the replacement NFOs train in operational fleet aircraft with student replacement pilots under the guidance of experienced instructors.

CAREER DEVELOPMENT

The First Sea Tour

After graduating from the training command and the FRS, the junior aviator's career as a naval aviator begins in earnest. From this point on, the responsibilities and checkpoints in pilot and NFO career development are essentially the same. This development is depicted in figure 6-1 and explained in detail in the paragraphs to follow. The numbers at the left of the diagram represent the number of years of commissioned service at which time various events are likely to occur. It is important to stress that the career pattern depicted in figure 6-1 represents only a general progression. While it is true that successful aviators will by the end of their careers have completed most of the steps, the order and timing of these steps differs from person to person. It must also be pointed out that completion of the steps outlined in this diagram in no way assures success, nor does pattern alteration preclude success.

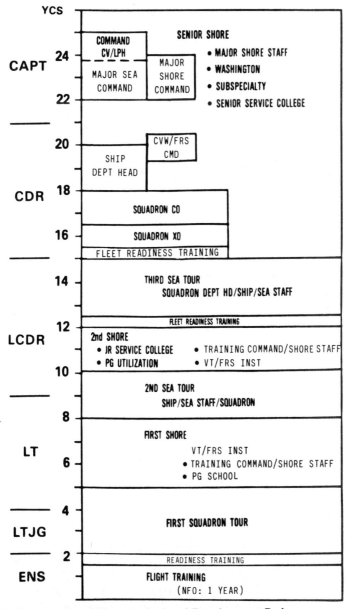

Fig. 6-1. Aviation Officer Professional Development Path

The initial squadron tour lasts approximately three years. This is a time for the aviator to learn while contributing to the attainment of the squadron's mission. After reporting to the new squadron, an officer can be assigned as a branch or division officer. In the squadron organization this is a starting point. The branch officer reports to the division officer who reports to the department head, who is responsible to the executive officer and the CO.

The duties of branch officer are the primary duties of the junior officer. How well these responsibilities are performed will determine the largest portion of an officer's fitness report. Professionalism in the air is demanded of all aviation officers and is also reflected in the fitness report.

Most junior aviation officers will have the opportunity to serve in more than one of the squadron's departments during their first tour. The four departments in most squadrons are operations, maintenance, safety, and administration. There are good, challenging jobs in each of these departments. During the first squadron tour the junior officer will gain experience in leadership, in the daily routine of the squadron, and in learning to employ and utilize personnel.

The First Shore Tour

Once the officer has completed the first squadron tour, he becomes available for assignment to a shore duty billet. These billets, many and varied, are located in every area where the navy has stations and facilities as well as in many predominantly civilian areas.

Because of the large demands of the aviation training community, many aviators will be assigned to the aviation training command to train the replacement pilots and NFOs needed to fill fleet seats. Others will be available to pursue advanced education opportunities, such as those offered at the Naval Postgraduate School in Monterey, California. Since there are senior billets in the navy that require graduate education, there is a strong need to develop a certain number of officers while they are at junior levels. Opportunity to attend the graduate school is open to 35 to 40 percent of aviators, and approximately 20 percent of those selected will be assigned upon completion of their first sea tour.

Officers assigned to Monterey for their first shore tour are primarily slated for the technical courses. After completion of graduate-level education, the officer will be assigned a subspecialty code that ensures one or more tours in the area of subspecialty. The graduate can expect to be detailed to a subspecialty billet on the next shore tour; if the subspecialty is one that can be employed at sea, he may be detailed to it on the next sea tour.

The Second Sea Tour

The aviation career pattern departs at this point from what has been the norm for many senior aviators in the navy. While aviators once received three operational squadron tours prior to command, that is less often the case now. The size of the officer's year group within aviation warfare subcommunities and the number of flying seats available determine whether an officer will have three such tours before command. For instance, if there is a shortage of F-14 pilots and NFOs going back for their department-head tours (the third sea tour), some pilots and NFOs qualified to fly F-14s and rotating back to their second sea tours will be ordered back to operational squadrons to make up the deficit. Conversely, if, as has been the case in the VP community during recent years, there are sufficient numbers of pilots and NFOs going back for their third sea tours, those in the VP community can expect a tour of sea duty in a nonsquadron-related billet. Shipboard tours, duty with afloat staffs, and oversea assignments fall into this category. The shipboard billets include catapult officer, carrier air-traffic-control officer, assistant strike operations officer, and assistant navigator. For helicopter pilots, these billets include aircraft handler or flight-deck officer on an amphibious-assault ship or air officer on an amphibious-transport dock. Officers who are assigned to a nonsquadron sea tour can expect a squadron tour for their third sea-duty tour.

The Second Shore Tour

The second shore tour for most aviation officers lasts from twelve months to three years, depending on the time available prior to entry into the department head tour. Ideally, the officer commences the department-head tour (third sea tour) approximately one and a half years before the first opportunity arises for aviation command screen. Accordingly, the length of the second shore tour may need to be adjusted to ensure that the milestone is met. Many of the assignment opportunities will be similar to those of the first shore tour; however, they will involve increased responsibility. The officer will again have the opportunity for graduate education with subspecialty utilization similar to that of the first shore tour. At this time, the officer will also become eligible for selection to the junior service colleges. These include the Naval War College as well as the junior service colleges of the army and air force. Selection for assignment to the Armed Forces Staff College in Norfolk, Virginia, is also a possibility.

The Third Sea Tour

The third sea tour is one of the most important tours for aviation officers. Most frequently this tour is also the second squadron tour, where the officer will have the opportunity to serve as a squadron department head. Those officers who have previously completed the second squadron tour should expect assignment to an aircraft carrier or deploying staff. Among the billets available are aircraft-handling officer or assistant air-operations officer on carriers and on staffs, air-wing-operations officer, flag secretary, and general-weapons and strike-operations officer. For helicopter pilots, these billets include assistant operations officer or assistant air officer on an amphibious-assault ship, or air antisubmarine officer on a cruiser-destroyer-group staff. The successful execution of the responsibilities of a lieutenant commander is the final test of leadership and managerial skills. The fitness reports received during this tour will be considered very carefully by the aviation command screen board.

AVIATION COMMAND SCREEN

The aviation command screen board is headed by a flag officer and is comprised of a cross-section of senior aviation officers who have commanded aviation squadrons. The board considers all aviation officers for operational and special-mission aviation squadron command. An officer is generally in the third sea tour during the period in which his record appears before the command screen board. The command opportunity for aviation officers is currently established at approximately 45 percent. This means that 45 percent of all pilots and NFOs serving in grade as commanders will have a command of some type. This includes force support and training squadrons. The opportunity to command an operational squadron is more restricted. An officer can generally assume that those communities with larger numbers of eligible officers will offer less of an operational command opportunity, while those with fewer eligible officers, such as single-seat squadrons, will offer higher operational command opportunity. The command tour that includes initial assignment as executive officer will normally begin one to two years after selection. Squadron-command tours, including time spent as an executive officer, are generally thirty to thirty-six months.

CAPTAIN-MAJOR COMMAND

Aviation-warfare captains are screened by a formal board for major sea command, major shore command, and as candidates for major project manager. The annual screening board, which is made up of flag officers, selects the best qualified aviation captains from each promotional year group.

Major sea commands for aviation captains consist of both amphibious and mobile-logistics force ships and patrol air wings. Approximately half of those aviator captains assigned to a major sea command will serve a sequential sea command in units such as amphibious and service force squadrons, amphibious-assault ships, and aircraft carriers.

OPNAVINST 5450.205A lists those shore commands designated by the CNO as major shore commands. The length of a tour is twenty-four months. Current policy precludes an officer from being assigned to both a major sea and a major shore command.

For aviation-warfare captains, the length of a major sea command will be two years. Those selected for second sequential sea commands, however, will have their major command for fifteen months and their sequential sea command for eighteen months. Command of a nuclear surface ship is set at three years.

FEMALE AVIATOR CAREER PLANNING

The career pattern for female aviators has been developed to parallel that of their male counterparts, subject to the assignment restrictions imposed by Title 10 USC, Section 6015. This legislation precludes assignment of women to squadrons with a combat mission. The female aviator will follow the same sea-shore rotation cycle, including tour lengths, as her male contemporaries. Ashore, the female aviator may be assigned any 13XX billet for which she is qualified by grade and experience.

TRAINING DURING SHORE TOURS

There are several other types of training available to pilots and NFOs: postgraduate school, service college, the U.S. Naval Test Pilot School, and the British Empire Test Pilot School.

Female aviators follow a career pattern similar to that of their male counterparts but by law cannot be assigned to combat squadrons. (U.S. Navy)

POSTGRADUATE TRAINING

The ever-increasing technical complexity of naval weapons systems is particularly evident in naval aviation. The navy's postgraduate training program is an excellent means of acquiring advanced knowledge and training for officers who want to become specialists as well as for those who intend to remain "in the line," using their new skills as secondary specialties. The Naval Postgraduate School in Monterey provides an opportunity for qualified officers to continue their education while on active duty and drawing full pay

The Naval Postgraduate School in Monterey, California, provides an opportunity for pilots and NFOs to continue their education and complete requirements for advanced degrees. (U.S. Navy)

and allowances. The wide range of postgraduate subjects reflects the navy's need for officers with advanced education in diversified areas.

Details of the fully funded programs are promulgated annually in OPNAVNOTE 1520.

In addition to postgraduate school, there are other advanced academic programs open to officers. The courses of primary interest to naval aviation officers are in the following areas: aeronautical engineering, ordnance engineering, meteorological engineering, electronics and communications engineering, and operational analysis. These courses lead to an M.S. degree in the field of specialization after two or three years of intensive study. It is assumed that the officers entering will have had training in mathematics and science equivalent to that offered at the U.S. Naval Academy. These are not additional training courses but extensions of previous educational programs; as such, they are attractive to all aviation officers qualified for entrance and interested in advancement in the technical fields.

In a normal career path, the best time to enter the postgraduate program is during the tour of shore duty following the first tour of duty with the fleet. This allows the officer to complete his three years of education without upsetting his basic rotation pattern and also returns him to the subspecialty environment before too many years have passed.

No special letters of application are required, but an officer should indicate his preferred postgraduate field on his officer preference card.

SERVICE COLLEGES

A number of service-college courses are available for which application is not required. Enrollment is by action of an annual selection board, which considers all available naval officers within the eligibility zone. The colleges are the Naval War College, Newport, Rhode Island; the Armed Forces Staff College, Norfolk, Virginia; the Industrial College of the Armed Forces, Washington, D.C.; and the National War College, Washington, D.C.

In addition, the Defense Language Institute, with branches in

Washington, D.C., and Monterey, California, offers courses in approximately fifty languages at military training activities, schools of government departments, civilian universities, and commercial language schools. A language aptitude test is required prior to selection.

TEST PILOT SCHOOLS

If a person is interested in getting into the test pilot field, there are three schools available: the U.S. Naval Test Pilot School, the USAF Aerospace Research Pilot School, and the British Empire Test Pilot School. Candidates for these schools are selected twice a year by a board convened by the commander of the NMPC.

U.S. Naval Test Pilot School

To enter the U.S. Naval Test Pilot School, located at the naval air station, Patuxent River, Maryland, an officer must be a pilot or NFO in the grade of lieutenant (j.g.), lieutenant, or lieutenant commander. College-level algebra and physics are prerequisites, and calculus and two years of college engineering are helpful. Normally, fifteen hundred hours of flight time is considered the minimum for applicants, although exceptions have been made.

The mission of this school is to train highly motivated and experienced fleet pilots and NFOs to become fully qualified naval test pilots or naval test flight officers. Assignment to this school is highly prized by young naval officers. Upon graduation, pilots and NFOs complete a normal tour of shore duty at the Naval Air Test Center, Patuxent River, Maryland, while serving as project pilots and NFOs who test and evaluate new aircraft and aircraft components.

The eleven-month course of instruction consists of two phases. The academic phase includes aerodynamics, aircraft and engine performance, and related aeronautical engineering subjects. During the flight phase an officer learns planning, more about flying, and how to report test flights in advanced airplanes. He becomes familiarized with new types of aircraft, instruments, and night proficiency flying. Both pilots and NFOs average about twenty flying hours per month.

USAF Aerospace Research Pilot School

Presently the navy has a quota for naval aviators who attend the USAF Aerospace Research Pilot School, which starts each February and August at Edwards Air Force Base in California. The mission of the school is to train pilots in the latest methods of testing and evaluating aircraft, manned space vehicles, and related aerospace equipment.

To be selected, you must have a bachelor's degree in engineering, physical science, or mathematics, and you must have acquired a minimum of one thousand hours of total pilot time and five hundred hours of jet time.

British Empire Test Pilot School

This school, located in Farnborough, England, trains pilots who test-fly British Ministry of Supply experimental aircraft. Two U.S. naval aviators are selected each year for the ten-month course, which begins in February. Applicants must have completed college algebra and physics and have recent operational pilot experience.

QUALIFICATION CRITERIA FOR AVIATION-CAREER-INCENTIVE PAY

One additional point that should be considered when planning an aviation career is the requirement establishing eligibility for continuation of aviation-career-incentive pay (ACIP). At certain checkpoints the aviation officer must have achieved a prescribed ratio of operational flying time to total aviation service. These checkpoints are known as gates. To meet the first gate, he must have served as an officer in an operational flying billet for six of the first twelve years of aviation service (including flight training). Upon completion of the six years of operational flying, he is entitled to receive continuous ACIP through the eighteenth year of aviation service, regardless of the billet to which he is assigned. If he fails to complete six of his first twelve years in an operational flying billet, he will receive ACIP only when he is actually filling an operational flying billet.

The second "gate" is measured at the completion of eighteen years of aviation service. If, at this time, an aviation officer has accumulated eleven years of operational flying, he is entitled to ACIP through the twenty-fifth year of commissioned service. If he has completed only nine years of operational flying at the eighteen-year gate, he will receive ACIP through the twenty-second year of commissioned service. Completion of less than nine years of operational flying in the first eighteen years of aviation service authorizes him to receive ACIP only when he is filling an operational flying billet.

Since ACIP is based on completion of operational flying time, it is important to discuss briefly how this operational flying time can be obtained. All sea-duty squadrons are composed of 100 percent operational flying billets for pilots and NFOs. Also, the time spent in the training command and the FRS will be credited as operational flying. Numerous billets on aviation ships (CVs/ LPHs) and seagoing staffs of attack-carrier air wings and carrier groups also qualify as operational flying billets in accruing "gate time." The majority of the operational shore-duty flying billets are located in the aviation training command and the FRS squadrons. Other shore commands that contain a high percentage of operational flying billets are the Naval Air Test Center, VX squadrons, and naval air stations.

REPORT ON THE FITNESS OF OFFICERS

Periodically the performance, qualifications, and potential of each officer will be evaluated by his CO or reporting senior in a written form known as a fitness report. His CO will evaluate his performance in primary assignments, such as maintenance division officer, as well as any collateral duties for which he may be responsible. His skill and airmanship will also be evaluated in the same report.

Fitness reports form the most vital part of an officer's record. They receive careful consideration by promotion boards, selection boards for special programs (such as postgraduate education and augmentation), and officer detailers when an officer is considered for new duty assignments.

Normally, fitness reports will reflect an officer's ability to plan and manage assignments, supervise subordinates, work in harmony with others, support the navy, remain composed under stress, and advance equal-opportunity goals. Specific personal traits, such as judgment, initiative, forcefulness, behavior, and military bearing, are also rated.

Special skills and qualifications are also carefully evaluated so that the most skilled aviation officers will be chosen for the most demanding flying assignments.

SELECTION BOARD
INFORMATION

Selection boards are a mystery to most officers. However, if they knew the thoroughness and impartiality of the selection board procedure, their confidence in the navy's selection system would be enhanced.

The composition of selection boards is regulated by law. The selection of members is, of course, extremely important. The boards are invariably composed of the top talent available in the navy and are carefully selected to give a wide range of experience and diversity of background. The composition of each board is reviewed on the highest level and finally approved by the secretary of the navy. At first the names of board members are kept confidential so that they will not be specially influenced in advance.

When a selection board first meets in Washington, the chief of naval personnel and members of his staff make a series of presentations. These cover the needs of the navy as a whole for officers in the grade under consideration, the various laws and regulations affecting promotions, the officer grade structure, and distribution requirements. Career patterns are discussed in the light of the navy's present and future requirements, and board members are told that there are no fixed career patterns, that there are many avenues to success. These briefings, while oriented towards the needs of the navy, are intentionally broad in scope. There is no attempt to establish finite guidelines or to dictate who or what type of person should be selected: This is the responsibility of the board. It is made clear to the board that they are free to choose any rule and procedure they wish within the scope of the precept, which is a letter to the president of the board from the secretary of the navy.

After the briefing by the chief of naval personnel and his staff, the board convenes in the board room. The members and the recorders, who give their oath to perform their duties to the best of their abilities, are sworn in. The precept is then read. It tells the board which officers are in the promotion zone, it indicates the numbers of eligible officers the board may look at above the zone, and it further specifies the number of people who can be selected for "head and shoulders" early selection.

After the board has formally convened, members sit in executive session to tackle the most important problem before them. They are charged with the solemn responsibility of selecting, to the best of their collective ability, those officers who have the greatest potential for service in the next grade. The board deliberates at great length over the criteria to be used in selection. They ask themselves many penetrating questions before laying down specific criteria.

Several selection boards will look at each officer's record before he is passed over for good. No officer is permitted to serve on two consecutive selection boards for the same grade.

Of the many thousands who have examined the navy's promotion system, most agree that no service or organization has a better, fairer selection system than the one the navy has evolved. It is not perfect; however, we can indeed be thankful that our selection system is not affected by nepotism, by a colleague who marries the boss's daughter, by the amount of stock someone owns in a company, or by superior talent hired from another company.

An officer's strengths are highlighted so that selection boards and other officers responsible for his career development know his character and potential. A CO normally shows a subordinate his fitness report, and after discussion of it, the officer is able to develop his strengths further and devote extra effort to weak areas.

An officer's performance is evaluated and compared with the performances of those officers who have served for approximately the same length of time and who have similar experience. Therefore, a junior officer does not have to have served a lengthy period to receive top marks—if he does his best his performance will be recognized. Success as a naval officer and as a naval aviator is achieved by superior performance. Each assignment will be "career enhancing" if it is done well.

7

Command Organization

Revised by R. H. Thompson

The various commands and units of the navy can be divided into two major categories known as the operating forces and the shore establishment. This chapter is concerned primarily with naval aviation as a part of the former.

THE OPERATING FORCES

The operating forces of the navy consist of fleets, seagoing forces, the Military Sealift Command, and certain shore activities. They are responsible for the conduct of naval operations as the navy's mission is carried out in the pursuit of national policy and interests.

Command of the operating forces is exercised via two chains, administrative and operational. The administrative chain runs from the secretary of the navy and the CNO to the operating forces. In the operational chain, command flows from the president through a commander of a unified or specified command such as commander in chief, Atlantic (CINCLANT), or commander in chief, Pacific (CINCPAC), and thence to an operational commander such as commander in chief, Atlantic Fleet (CINCLANTFLT) or commander in chief, Pacific Fleet (CINCPACFLT). It should be noted here that CINCLANT and CINCLANTFLT are headed by the same individual.

The administrative organization is permanent, while the operational side is task oriented and can be restructured to meet changing operational requirements.

TYPE COMMANDERS

Most aviation commands such as aircraft carriers, carrier air wings, patrol wings, squadrons, etc., come under the administrative command of an aviation type commander, COMNAVAIRLANT or COMNAVAIRPAC (commander, Naval Air Forces, Atlantic and Pacific, respectively). A few do not. Some air test and evaluation squadrons (VXs), for example, serve under commander, operational test and evaluation force (COMOPTEVFOR). Amphibious assault ships, while primarily involved in helicopter operations, are under the operational control of commander, Surface Force (Atlantic or Pacific), as are tactical support squadrons.

COMNAVAIRLANT and COMNAVAIRPAC command-assigned aircraft carriers, squadrons, and shore activities exercise operational control over such forces and maintain all units in optimum states of readiness. They standardize aviation operations, maintenance, and administrative procedures, and recommend aviation policy to fleet commanders on questions concerning the organization, maintenance, distribution, and employment of fleet aviation. They are responsible for the flight training of air wings, squadrons, and units attached to ships undergoing shakedown or refresher training. The aviation type commander's staff includes eight main divisions: administrative, supply, communications, material, readiness, comptroller, personnel, and safety.

Functional wing commanders are subordinate to aviation type commanders, who command similar air units or related commands. Commander, Tactical Wings, Atlantic, for example, includes the early-warning wing, the light-attack wing, the medium-attack wing, and the fighter wing as well as the air stations where these wings are based. Functional wing commanders perform a type com-

Commander, Naval Forces, Atlantic, is a type commander with headquarters in Norfolk, Virginia. (U.S. Navy)

mander's functions for assigned aviation units, coordinate aviation matters, and aviation logistic support in their areas, and advise the type commander regarding naval air force and shore activity matters in their areas. They administer their assigned aviation squadrons and commands and distribute and assign personnel; administer, control, and coordinate aviation operations, training, and inspections; exercise cognizance over both aviation material and aircraft maintenance; supervise aviation supply for their assigned units and in their assigned areas. Functional wing commanders also have medical, communications, intelligence, and public works functions.

FLEETS

The navy has four numbered fleets, the Third and Seventh in the Pacific, the Second and Sixth in the Atlantic. (Generally speaking, the Pacific Fleet has "odd" numbers, and the Atlantic Fleet "even" numbers, for fleets and carrier groups.) The U.S. Sixth Fleet, permanently deployed to the Mediterranean, has a NATO title, Naval Striking and Support Forces South. The U.S. Seventh Fleet is permanently deployed in the western Pacific. Ships are rotated to and from these fleets for deployments, which usually last from six to seven months. The Third Fleet is a roving Pacific Ocean fleet, which has responsibility for the eastern half of the Pacific. The Second Fleet is a roving Atlantic Ocean fleet with the NATO title, Striking Fleet Atlantic. Both the Third and the Second Fleet have overall operational scheduling and training responsibilities for Pacific and Atlantic ships not assigned to other commands.

FORCES

Under the fleet commanders are certain forces. The Sixth Fleet, for example, has a battle force (CTF-60), an amphibious force (CTF-61), a landing force (CTF-62), a service force (CTF-63), an FBM submarine force (CTF-64), an area antisubmarine force (CTF-66), a maritime surveillance and reconnaissance force (CTF-67), a special-operations force (CTF-68), and an attack-submarine force (CTF-69).

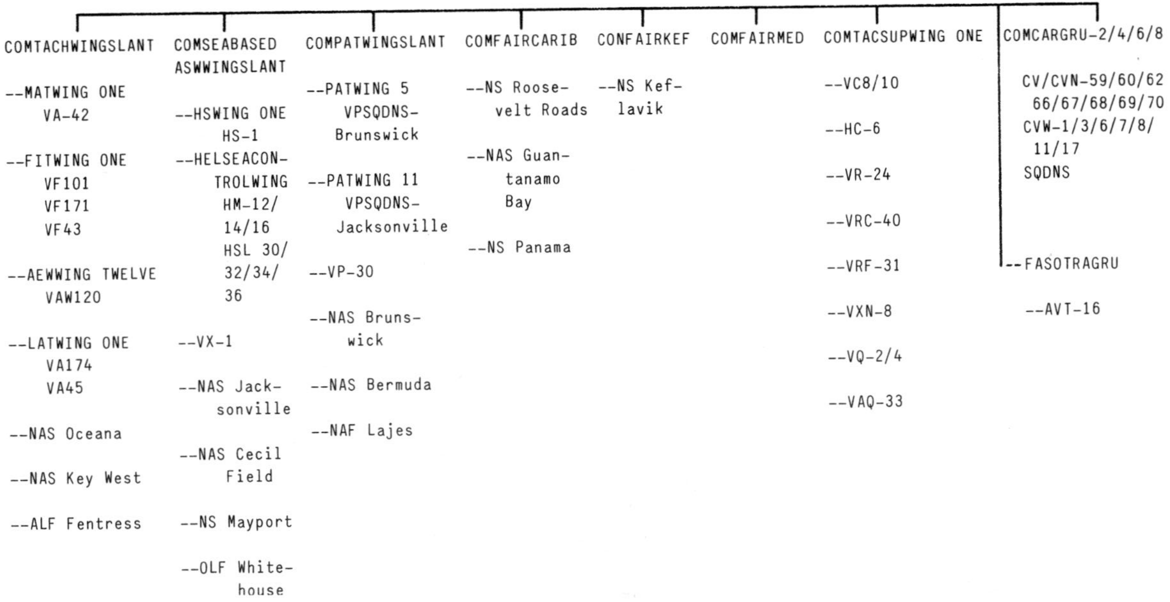

COMTACHWINGSLANT	COMSEABASED ASWWINGSLANT	COMPATWINGSLANT	COMFAIRCARIB	CONFAIRKEF	COMFAIRMED	COMTACSUPWING ONE	COMCARGRU-2/4/6/8
--MATWING ONE VA-42	--HSWING ONE HS-1	--PATWING 5 VPSQDNS- Brunswick	--NS Roose- velt Roads	--NS Kef- lavik		--VC8/10	CV/CVN-59/60/62 66/67/68/69/70 CVW-1/3/6/7/8/ 11/17 SQDNS
--FITWING ONE VF101 VF171 VF43	--HELSEACON- TROLWING HM-12/ 14/16 HSL 30/ 32/34/ 36	--PATWING 11 VPSQDNS- Jacksonville	--NAS Guan- tanamo Bay			--HC-6 --VR-24 --VRC-40	--FASOTRAGRU
--AEWWING TWELVE VAW120		--VP-30	--NS Panama			--VRF-31	--AVT-16
--LATWING ONE VA174 VA45	--VX-1 --NAS Jack- sonville	--NAS Bruns- wick --NAS Bermuda				--VXN-8 --VQ-2/4 --VAQ-33	
--NAS Oceana	--NAS Cecil Field	--NAF Lajes					
--NAS Key West	--NS Mayport						
--ALF Fentress	--OLF White- house						

Fig. 7-1. COMNAVAIRLANT

COMFITAEWWINGPAC	COMLATWINGPAC	COMMATVAQWINGPAC	COMASWWINGPAC	COMPATWINGPAC	COMFAIRWESTPAC	COMCARGRU	FASOTRAGRU
--VF125	--VA-122	--VA128	--VS-41	--PATWING-1 SQDNS	--VC5	--COMCAR GRU1/3/ 5/7	
--VF126	--VFA-125	--VAQ129	--HC-1/3/7/11	--PATWING-2 SQDNS	--VRC30		
--VC13	--VA127	--NAS Whidbey	--HS SQDNS	--NAS Moffett	--NAS Cubi Point	--CV/CVN41 43/61/ 63/64/ 65	
--VX-4	--VX-5	--VAQ Sqdns	--HSL SQDNS		--NAS Agana		
--VAW-110	--NAS Lemoore		--NAS North Island	--NS Adak	--NAF Atsugi	--CVW SQDNS 2/5/9/11/ 14/15	
--NAVFITWEPSCOL	--NAS Alameda		--ALF San Clemente	--VP31	--NAF Misawa		
--NAS Miramar	--NAS Fallon		--OLF Imperial Beach	--NAF Midway Island	--NAF Diego Garcia		
--NAF El Centro				--NAS Barbers Point			
				--ALF Crows Landing			
				--PATWING-10 SQDNS			

Fig. 7-2. COMNAVAIRPAC

BATTLE FORCE ORGANIZATION

The major offensive element of the navy is the battle force. In common with other task forces, it is a fluid, semipermanent organization composed of various types of ships and aircraft. It fluctuates in size as ships join its command, are detached for other operations, or are sent back to bases for overhaul or repair. It may be reconstituted from within or subdivided to meet various operational requirements.

A basic tactical unit of the battle force is the carrier battle group, which consists of one or more carriers supported by missile cruisers and screened by destroyers and attack submarines.

The modern CV/CVN is capable of adapting to the wide range of scenarios and threats it is exposed to. A carrier air wing can be altered slightly to further optimize the carrier's capability.

Any combination of air, surface, and subsurface assets can be further organized as battle group(s) and employed as needed for action against enemy surface ships, submarines, or coastal instal-

The carrier battle group is a tactical unit centered around one or more aircraft carriers and supported by combatants and other ships. (U.S. Navy)

lations. A battle group may be one of several in the task organization of a battle force.

Replenishment ships may be assigned directly to the attack carrier striking force and organized as an underway replenishment group under the control of the striking force commander, or they may be organized as part of an underway replenishment force, servicing several task forces and protected by a sea group.

On reporting to the fleet, carriers are assigned to a group commander. He can be either a carrier-group commander or a cruiser-destroyer-group commander. Group commanders function in both an operational and an administrative capacity and as part of the task fleet or force. They are responsible for the operational planning and coordination of the air, surface, and subsurface operations of their assigned ships and aircraft as subordinates of the air/surface type commanders. They are also responsible for the readiness, training, and administration of assigned ships and embarked aircraft. The eight carrier groups are 2, 4, 6, and 8 in the Atlantic; 1, 3, 5, and 7 in the Pacific.

The carrier is under the administrative and operational control of the appropriate carrier-group or cruiser-destroyer-group commander. The carrier CO is responsible for operational training, readiness, the conduct of operations, and maintaining the embarked air wing as an integrated unit of the ship. Carrier operations are described in the following publications:

1. CVN/CV NATOPS Manual
2. NWP (Naval Warfare Publication) 8-1, Composite Warfare Doctrine
3. NWP 20, Striking Force Operations
4. NWP 21, Battle Group ASUW
5. NWP 51, Air Tactics for Sea Control
6. NWP 52, Air Tactics for Power Projection

FLEET AIR COMMANDERS

Fleet air commanders coordinate the activities of type-commander fleet-aviation units assigned to remote geographical areas outside the United States. Such assignments are made to best suit the training and operational mission requirements of the commands or units concerned. While fleet air commanders normally report for additional duty only to the type commander, tasking and responsibilities very closely parallel those of the functional wing commanders. These responsibilities include assuming type-commander functions for units assigned within a particular geographic area; coordinating fleet-aviation matters and controlling aviation logistic support to fleet air units in the area of responsibility; advising the type commander and force commanders of corresponding areas regarding naval air forces and shore activities based or located in their areas; and sustaining combat/operational readiness of aviation units by providing a realistic training environment.

PATROL WINGS

A patrol wing is a mobile, operational, and administrative unit composed of multiengine land planes organized in two to six squadrons. In the Pacific Fleet, individual patrol wings are assigned to the fleet air commander in whose area the wing is based, but they are under the overall command of COMPATWINGSPAC (commander, patrol wings, Pacific). In the Atlantic, COMPATWINGSLANT (commander, patrol air wings, Atlantic) exercises logistic and type-command responsibilities for all Atlantic wings. Patrol-wing commanders are responsible for the administration, training, operational readiness, and inspection of assigned units. They also have cognizance over supply, material, aircraft maintenance, and in some cases, communications, intelligence, medical and dental functions.

OPNAV ORGANIZATION

The CNO directs the operation of the U.S. Navy from his Washington office, the Office of the Chief of Naval Operations (OPNAV), in the Pentagon. OPNAV is a vast, complex, constantly changing organization, responsible for handling a multitude of details concerning the world's largest navy and, at the same time, keeping abreast of rapid changes in ship, weapons, aircraft, and astronautics technology.

Because of its size, OPNAV often seems immobile and an undesirable organization to be assigned to. However, it is where all policy matters affecting the navy are made or resolved; it is a fascinating place to work and one should approach any opportunity for assignment there as a chance to put his fleet experience to work and carry his ideas forward to the highest echelons of the navy. The following paragraphs give a very brief description of OPNAV, principally of the air staff.

Chief of Naval Operations

The CNO (OP-00) is assisted by his principal and senior deputy, the vice chief of naval operations (VCNO, OP-09) and six deputy chiefs of naval operations (DCNOs):

OP-01   Manpower, personnel, and training
OP-02   Submarine warfare
OP-03   Surface warfare
OP-04   Logistics
OP-05   Air warfare
OP-06   Plans, policy, and operations

Within OPNAV, there are also directors of major staff offices:

OP-090   Director of navy program planning
OP-093   Surgeon general
OP-094   Director, command and control
OP-095   Director, naval warfare
OP-098   Director, research development technology & evaluation
OP-007   Chief of information
OP-008   Naval inspector general
OP-009   Director of navy intelligence

Since OP-05 is concerned with naval aviation, this office will be described in some detail (see figure 7-3). However, many naval aviators occupy key billets in the other offices within OPNAV.

DCNO (Air Warfare)

This is considered the top aviation job in the navy. OP-05 acts as the CNO's principal advisor on naval aviation matters, including air warfare, and as his representative in naval air operational matters involving other government and civil agencies. He implements the responsibilities of the CNO with respect to naval aviation pro-

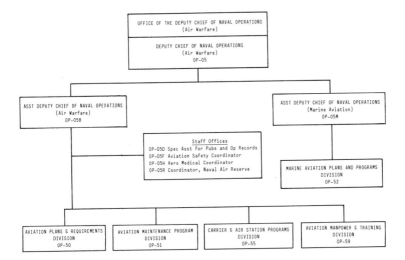

Fig. 7-3. Organization of OP-05

grams, including the naval air reserve, and to shipboard support requirements for aircraft carriers and certain aviation ships. To indicate a few of his functions, he:

1. Acts as the principal advisor to the CNO on, and establishes overall policy for, the preparation and conduct of air warfare.

2. Determines plans and requirements, states major characteristics, establishes programs, and acts as program sponsor for the development, procurement, modernization, and alteration of aircraft, aircraft weapon systems, and related equipment.

3. Puts together the budget for the procurement of naval aircraft and missiles and represents the naval aviation community before the secretary of defense and Congress.

4. Directs and controls the procurement of and logistic support for all naval aircraft.

5. With the director of naval warfare (OP-095), does force-level planning for aircraft and aircraft carriers to establish

goals and to propose acquisition and support strategies to achieve these goals.

6. Evaluates developments in aviation, analyzes, in conjunction with the DCNO (plans and policy), the capabilities of naval aviation with respect to strategic plans, and recommends new and more effective methods for utilizing naval aviation in collaboration with other forces.

7. Coordinates, measures, and promotes the readiness of the carrier forces in accordance with established air-warfare plans and programs.

8. Directs the development of and prepares criteria for determining naval aviation manpower requirements, and formulates policy on matters pertaining to manpower requirements and the allocation, distribution, performance, and qualifications of aviation personnel.

9. Directs and supervises flight and ground training conducted in functional air training commands.

10. Formulates and implements policy pertaining to flight rules and regulations and the utilization and control of naval aircraft, and develops requirements and establishes and monitors programs for air-traffic-control systems and aids to air navigation.

11. Directs the naval aviation safety program.

Assistant DCNO (Air Warfare)

OP-05B is the executive officer and principal advisor to the DCNO (air warfare).

Assistant DCNO (Marine Aviation)

OP-05M assists the DCNO (air warfare) to ensure that marine aviation plans and programs are in all respects adequate. He is "double hatted;" his primary position is deputy chief of staff (air), U.S. Marine Corps.

The principal assistants to OP-05 and OP-05B are in the offices of publications and operational records (OP-05D), safety (OP-

05F), the aeronautical medical coordinator (OP-05H), and the naval air reserve (OP-05R). Their missions and functions are self-explanatory.

Operating Divisions Under DCNO (Air Warfare)

There are five main "OPs" under the DCNO (air warfare): aviation plans and requirements (OP-50), aviation maintenance programs (OP-51), marine aviation plans and programs (OP-52), carrier and air station programs (OP-55), and aviation manpower and training (OP-59).

Aviation Plans and Requirements Division

This division prepares plans and defines requirements for the provision of naval aviation forces and their logistic support; coordinates and integrates these plans and requirements; and monitors joint, international, and navy plans, programs, and policies involving the mission of the DCNO (air warfare). Within OP-50 are the aircraft- and weapon-program coordinators who are responsible to the CNO for their programs (for example, the F/A-18 program, the F-14, and the P-3). Also in OP-50 are the aviation planners who recommend what and how many aircraft are procured and how they are distributed to the fleet. The budget shop for OP-05 is also in OP-50.

Aviation Programs Division

This division manages naval aviation maintenance, supply, and flying-hour programs. OP-51 is responsible for improving the readiness of naval aviation.

Marine Aviation Plans and Programs Division

This division formulates and coordinates plans and initiates action to fulfill the requirements of marine aviation in matters of organization, personnel, operational readiness, and logistics. It ensures that marine aviation is developed and supported in concert with the overall naval aviation program.

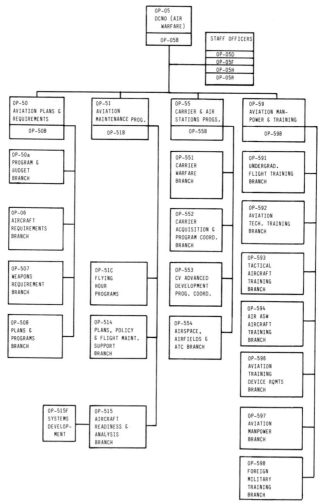

Fig. 7-4. OP-05 Organization Relationships

Carrier Programs Division

This division establishes carrier programs and policies for operational readiness, training, and requirements, and coordinates and directs all planning for the acquisition of carriers, the determination of shipboard requirements, budgeting, design, outfitting, and manning. It makes sure that all elements of the planning are consistent with each other and with navy and Defense Department guidelines. OP-55 is also responsible for the maintenance of naval air stations and facilities, the upkeep and modernization of the air-traffic-control system, and the interface with the Federal Aviation Administration.

Aviation Manpower and Training Division

This division formulates the aviation-manpower and training requirements of the navy. It establishes policy and directs the efforts of naval bureaus, offices, and commands to ensure that adequate training personnel are available when and where they are needed. It acts for the DCNO (air warfare) and advises him on the training and qualifications of naval aviation personnel and foreign military personnel.

8

Maintenance and Supply

Revised by Chief Warrant Officer J. Emmert, USN
and Lieutenant Peter Reynierse, USNR

The most highly skilled aircrew with the most modern and sophisticated equipment is ineffectual if its aircraft does not fly and
its weapons do not function properly. In U.S. naval aviation, a
well-conceived and finely tuned maintenance program backed by
a responsive aviation supply system ensures maximum availability
of aircraft and optimum performance of associated systems as a
matter of routine.

THE NAVAL AVIATION MAINTENANCE PROGRAM

OPNAVINST 4790.2 explains the naval aviation maintenance program (NAMP). The publication is divided into five volumes as
follows:

   I.   Concepts, Objectives, Policies, Organizations and
        Responsibilities
  II.   Organizational-Level Maintenance
 III.   Intermediate-Level Maintenance

IV.   Depot-Level Maintenance

V.   Data-Processing Requirements

The basic objective of NAMP is to achieve maximum material readiness in a safe and cost-effective manner. Aircraft maintenance is a responsibility of command, and a squadron is judged to a large extent on the effectiveness of its maintenance department and its ability to meet program objectives.

THE THREE LEVELS OF AVIATION MAINTENANCE

Depot-level maintenance refers to the work done in an industrial facility. It includes rework, overhaul, and major repair and modification of aircraft, components, and equipment. It also includes the manufacture of certain aeronautical parts and frequently the manufacture of kits for use in aircraft and equipment modification. Depot-level maintenance is carried out by contractors and by naval

This aircraft fuselage is being delivered to the naval air rework facility in Jacksonville, Florida, for depot-level maintenance. It will be reconstructed and delivered to the fleet as a new aircraft. (U.S. Navy)

air rework facilities located at naval air stations in Norfolk, Virginia; Jacksonville, Florida; Pensacola, Florida; North Island, California; and Alameda, California, and at the marine corps air station at Cherry Point, North Carolina. Depots also perform lower levels of maintenance.

Intermediate-level maintenance is performed in centrally located facilities both ashore and afloat for the support of operating units, stations, or ships. It may include:

—Repairing, testing, updating, modifying, or checking aircraft components
—Complete engine repair
—Module repair of electronic components
—Calibration/qualification of selected equipment
—Technical assistance to supported units
—Other appropriate services

Intermediate-level maintenance is performed on a jet engine at the aircraft maintenance department of the naval air facility in Washington, D.C. (U.S. Navy)

Ashore, intermediate-level maintenance is performed by aircraft maintenance departments of naval air stations, air facilities, and naval stations that support aircraft. On board carriers, it is carried out by aircraft intermediate maintenance departments, which are discussed in chapter 4.

Organizational-level maintenance is performed on assigned aircraft by squadrons, units, and detachments, and by the operations maintenance divisions of stations. It includes flight-line operations

Organizational-level maintenance is performed by squadron personnel. (U.S. Navy)

and servicing, daily and preflight inspections, minor adjustments in preparation for flight, periodic inspections of aircraft and equipment, repairs and adjustments that do not require extensive shop facilities. The organization of a squadron maintenance department is outlined in chapter 3.

THE 3M SYSTEM

Aircraft and their systems have become so complex that a comprehensive system was developed in the mid 1960s to provide efficient maintenance management. Known as the 3M system, it is really a combination of the planned maintenance system and the maintenance data system.

The planned maintenance system uses publications, charts, and cards to keep the maintenance process on schedule. These are known as the Periodic Maintenance Requirements Manual, sequence control charts, and maintenance requirements cards.

The maintenance data system of the 3M system provides to local and higher levels of management the tools needed to manage effectively the available assets of men, money, and material. All types of data are recorded on various forms (source documents) by maintenance and material personnel. To simplify and standardize the data and to make it compatible with automatic data-processing equipment, a series of alphanumeric codes have been developed. These codes are recorded on the source documents, verified by the supervisors, and processed daily by the local data-processing activity, ashore or afloat. Local daily and monthly management reports are given to the maintenance manager. Duplicate record files are made by the local data services facility and are mailed to the Navy Maintenance Support Office in Mechanicsburg, Pennsylvania, which is the central data-processing center for the navy. This office in turn compiles periodic or special reports containing worldwide data for use by local, intermediate, and top-level managers.

MAINTENANCE PERSONNEL AND TRAINING

The most important part of the aviation maintenance system is the people who make it work. Much time and effort is expended on the training of aviation maintenance personnel, the people who keep today's sophisticated systems on the line.

Key officers in a squadron maintenance department are the maintenance officer, assistant maintenance officer, and maintenance/material control officer. One of these officers is an aviation ground officer and one is a pilot or NFO in a flying status. If the squadron has a supply corps officer permanently attached, he is assigned to be the material control officer.

Aviation ground officers normally attend the sixteen-week aviation maintenance officer course at the naval air station in Pensacola. They also undergo training for a specific aircraft type in an FRS and may attend the three-week joint aviation-supply and maintenance-material management course at the navy supply school in Athens, Georgia.

Ideally, all aviation maintenance officers attend the sixteen-week aviation-maintenance-officer course. In any case, all attend a course conducted by a naval-air-maintenance training-group detachment within sixty days of assuming maintenance duties.

NATTC conducts a wide variety of technical training for both officer and enlisted personnel, as the following list indicates:

Class A—Provides training in basic technical skills for enlisted personnel
Class C—Provides advanced technical training for enlisted personnel
Class F—Provides team training and individual refresher training to both officer and enlisted personnel
Class P—Indoctrinates midshipmen, officer candidates, and newly commissioned officers
Class R—Indoctrinates enlisted personnel
Class V—Provides training leading to designation as a pilot or NFO

The Naval Air Maintenance Training Group provides on-site maintenance training for officers and enlisted personnel. Fleet Aviation Specialized Operational Training Groups instruct fleet personnel in the operational and tactical employment of specific equipment and systems. Naval air rework facilities conduct short courses in such things as troubleshooting, alignment, and specialized training in certain accessories and components. FRSs conduct FRAMP training for both aviation ground officers and enlisted maintenance personnel in the operation and maintenance of specific weapon systems or equipment.

The Department of Defense and the other military organizations offer courses to navy personnel. Additionally, the best maintenance personnel are occasionally sent to contractor plants for training on new weapon systems.

Available training programs are discussed in OPNAVINST 1500.11 (NOTAL) series instructions.

NAVAL AVIATION SUPPLY

Maintenance and supply are inseparable partners in the business of material readiness. In naval aviation, the Naval Air Systems Command joins with the Naval Supply Systems Command to keep

The Naval Aviation Supply Office in Philadelphia is the central control point of the naval aviation supply system. (U.S. Navy)

equipment, parts, and other material flowing to users at fleet and shore activities around the world.

THE AVIATION SUPPLY OFFICE

The Aviation Supply Office, located in Philadelphia, is the inventory control point for aeronautical parts. It is responsible for determining which items and how many are required, and for procuring, maintaining, and distributing such items.

Inventories of aeronautical material are held at naval supply centers, located near naval air rework facilities at operating air stations and at several major supply depots. The stocks of overseas depots fill shipboard and marine-air-group requirements and support other operating, maintenance, or industrial rework efforts in specific geographic areas. The status of the worldwide stock inventory is reported daily to the Aviation Supply Office so that accurate records of available material can be maintained.

Requirements are projected by the Aviation Supply Office based on operational and technical information from a number of sources.

Fig. 8-1. Naval Aviation Supply Distribution Points in the United States

The future aircraft operating program is designed by OPNAV, and additional details about maintenance and rework requirements are provided by the Naval Air Systems Command. Technical estimates of material needed for new aircraft and equipment programs are made by teams of technical personnel at a provisioning conference or by contractors. The need for material already in the supply system is determined from 3M and supply-system usage data. When new systems are provisioned, a "plan to maintain" is developed by teams consisting of naval aviators, aeronautical engineers, and supply-support personnel. The plan contains estimates of how often components will have to be repaired and replaced and of other support requirements. Data reported through the 3M program is used to revise procurement and support plans. Maintenance plans are revised according to requests from fleet and depot maintenance officers.

SUPPLY AT THE SQUADRON LEVEL

The basic unit for supply support of aircraft is generally the squadron. The squadron unit responsible for supply support is the material department or the material control division of the maintenance department. The maintenance/material-control division officer orders material for the maintenance of squadron aircraft, administers squadron funds, maintains custody of and is accountable for squadron property, and prepares reports.

Each squadron and aviation activity needs equipment, tools, and spare parts to perform its functions. Requirements are detailed in various allowance and outfitting lists published by the Aviation Supply Office. The lists are used by supply-support activities to determine the items, the repair, and the consumable parts needed to support their aircraft.

The Supply Support Center is the prime point of contact with the local supply department. Using procedures established by the 3M system, the Supply Support Center receives all requests for the direct support of aircraft from the squadrons by rapid communications and then delivers the materials at the earliest possible time.

A rotating pool of components is available for immediate supply support to maintenance organizations. Under this system, defective components receive intermediate-level maintenance and are replaced by components from the rotating pool. The pool is managed by the Supply Support Center.

NMCS (not-mission-capable supply) means that an aircraft is not operationally ready and that maintenance work cannot be performed because a part is unavailable. NMCS is the most urgent label that can be attached to a squadron's request for material. PMCS (partial-mission-capable supply) means that an aircraft has limited operational capability because a part is unavailable. Both PMCS and NMCS requirements are given command attention on a twenty-four-hour basis, not only by the air station supply department but by the local intermediate maintenance activity and the entire navy supply system as well. Every effort is made to procure or repair and deliver parts and materials in the shortest possible time.

FUNDING

In order to finance aircraft operations, a squadron receives from its fleet-air-force type commander a fund allocation commonly called the operating target (OPTAR). Each squadron is responsible for administering OPTAR funds and accounts to its functional wing and type commander.

Funds are sent directly to the intermediate maintenance activity to finance repair parts needed at the intermediate level.

A "flight packet" is normally issued to an aircraft commander for each flight made away from base. It is an official pocketbook and contains partially completed requisitions; purchase orders for use with commercial concerns and other government agencies; and other forms and instructions that may be needed for refueling, aircraft repair, or obtaining meals and billeting for the crew at other military or commercial airfields.

THE AIR-STATION-SUPPLY ORGANIZATION

The supply officer of a naval air station is responsible principally for providing the parts and materials for operating, maintaining, and repairing the aircraft based at his station. The air-station-supply department also publishes a series of instructions and notices and maintains liaison with squadron-maintenance/material-control officers to ensure a smooth flow of information and material.

SHIPBOARD-SUPPLY ORGANIZATION

Like the supply officer of an air station, the supply officer of a ship must provide the parts and materials needed to operate, maintain, and repair the aircraft on board. He is also responsible for food service, onboard retail stores, and disbursing. In the shipboard-supply organization, the unit primarily responsible for aircraft support is the aviation stores division. Much like the air-station-supply organization, the aviation stores division handles both aircraft materials and paper work, and acts as the squadron-ship liaison on all matters pertaining to aviation supply support. Here, too, personal contact between the squadron-maintenance/material-control officer, the ship's supply officer, and the aviation-stores-division officer guarantees optimum service.

An aviation-consolidated-allowance listing (AVCAL) is pre-

Personnel at the Naval Aviation Supply Office in Philadelphia supply the fleet with a steady flow of aeronautical parts. (U.S. Navy)

pared by the Aviation Supply Office between a ship's deployments. An AVCAL is based on the aircraft-equipment configuration lists provided by the fleet-air-force type commander in coordination with the squadrons preparing for deployment. Simply stated, the AVCAL is a compilation of all the individual outfitting lists applicable to the aircraft being assigned to the ship. An important element of a squadron's predeployment liaison with the ship's supply department is making sure parts have been obtained for late configuration changes and recommending any modifications.

THE AVIATION SUPPLY OFFICER

Supply officers specializing in aviation logistics have been around for a long time. Formal recognition of their special brand of expertise was made in 1984 when the CNO approved a plan to establish a naval aviation supply officers program with breast insignia for those who meet the qualifications.

Candidates who want to become naval aviation supply officers must complete a comprehensive program of approximately 350 hours of study and practical experience. They must also pass an oral examination administered by experienced supply and maintenance officers at their operating sites. Only those supply officers serving in designated aviation-supply billets or who have been detached from such billets within the previous two years are eligible to participate in this program.

Detailed information is provided in OPNAVINST 1542.4 and NAVSUPPUB (naval supply publication) 550.

Aviation Supply Officer Wings

MATERIAL READINESS

Material readiness requires a continuous stream of parts and components from industry through the supply system to carriers and air stations around the world. The Aviation Supply Office and its subordinate activities are the vital links in the logistics chain that makes the system work. An effective maintenance program supported by a dedicated and responsive supply system keeps squadron aircraft on the line and ready for any contingency.

Osborn

9

Safety and NATOPS

By Captain Rosario Rausa, USNR

"Every man and woman in the Navy is a safety manager!"
"Know your emergency procedures!"
"Safety saves!"

The slogans above would become platitudes in the world of naval aviation if the messages they conveyed weren't so important. From the outset in the flight training program at the naval air station in Pensacola, safety is emphasized, and eventually it becomes a daily goal of every navy and marine corps flyer.

ACCIDENTS ARE COSTLY

As of this writing, the U.S. Navy's aviation mishap rate is the lowest it has ever been. Compared with activities during World War II, when planes fell from the sky at an alarming rate, today's flight activities are, indeed, safe. Nevertheless, too many pilots and aircrew, men *and* women, are lost each year. Even though

Safety and NATOPS (Artist: Robert Osborn)

the declining rate of accidents is a welcome reflection of the continuing attention placed on safety, there can be no slackening of precaution. The cost in lives and in dollars spent for aircraft, training, and maintenance is simply too high. F/A-18 Hornets are worth twenty-five million dollars apiece. It takes three hundred thousand dollars or more these days to train a pilot, and the price tag rises with each passing year. The cost of human life is, of course, incalculable.

So for fledglings and veterans alike, including flag officers at the upper echelons of command, safety considerations are critical and second only in importance, perhaps, to actual operations.

THE NAVAL AVIATION SAFETY PROGRAM

The naval aviation safety program is described in OPNAVINST 3750.6N. Its chief purpose is to preserve human life and material resources by eliminating hazards. All aviation personnel should be familiar with its directives. Normally, when an officer checks into a squadron, his indoctrination briefings include reviews of pertinent safety instructions and safety "rules of the road" for operations in the air and on the ground. There are three main functions of a successful aviation safety program:

—Hazard detection
—Hazard elimination
—Safety information management

A naval aircraft mishap is defined as an unplanned event or series of events, directly involving naval aircraft, which results in injury or property damage exceeding ten thousand dollars. A class A mishap is one in which the total cost of property damage or injury is greater than five hundred thousand dollars and/or in which a person receives a fatal or permanently disabling injury. If the damage and/or injury costs are between one hundred and five hundred thousand dollars, the mishap is in the class B category. Class C accidents cost between ten and one hundred thousand dollars.

The main injury classifications are:

—Alfa, which signifies fatal injury
—Bravo, which represents permanent total disability

—Charlie, which means permanent partial disability
Delta through Golf injuries are of descending severity.

A common acronym in safety parlance is AMB—aircraft mishap board. An AMB consists of several officers, including a flight surgeon, and is convened to examine thoroughly an accident and to record its findings in a mishap investigation report to be submitted to higher authority.

If disciplinary measures were taken against those who provide information to an AMB, the flow of critical facts and observations from witnesses or those involved in mishaps would be impeded and the value of an investigation reduced. The primary purpose of an investigation is to prevent or reduce the likelihood of future accidents. Therefore the information supplied by witnesses and participants is confidential. It cannot be revealed to the public or to sources other than designated officials.

Each aviation squadron has an aviation safety officer (ASO). The ASO is the CO's principal advisor on all aviation safety matters. The ASO is usually a lieutenant or lieutenant commander on a second tour of flying duty who has graduated from the seven-week ASO course at the Naval Postgraduate School in Monterey.

The ASO enjoys department-head status and thus has direct access to the CO. This reflects the importance both the navy and individual units place on safety. Detachments or units smaller than squadrons may not have a full-time ASO. At least one officer will have ASO responsibilities on a collateral-duty basis, however. Aircraft carriers have full-time safety officers with department-head status, as do other larger commands and staffs.

THE NAVAL SAFETY CENTER

The U.S. Naval Safety Center, located at the naval air station in Norfolk, is commanded by a flag officer. Its mission is to collect, evaluate, and disseminate safety information. It develops aviation, submarine, surface ship, and shore safety programs. The aviation division is further divided into aircraft and facilities operations, maintenance and material activities, mishap investigations, and aeromedical matters.

The safety center puts out several publications, familiar items

The U.S. Naval Safety Center

in virtually all ready rooms in a ship and at shore commands throughout the navy and marine corps aviation community.

SAFETY INFORMATION

Approach magazine contains feature articles and accounts of accidents told by the personnel involved or qualified professionals.

Weekly Summary is a two-page pamphlet that lists the mishaps that have occurred in a particular week in commands the world over. It also contains short articles of timely importance.

Mech/Maintenance Crossfeed is a bimonthly magazine with articles on hazards, policies, and equipment pertinent to readiness and safety in aviation maintenance at all levels.

Escape Summary, published annually, documents ejections and over-the-side bailouts that have occurred during the year. It includes information concerning ejection methods used, injuries sustained, and the length of time required in rescue efforts.

ANYMOUSE

For navy and marine corps flyers and their support crews the term *Anymouse* has nothing to do with creatures from the animal kingdom. It is a comic substitute for the word *anonymous*. Anymouse reports are submitted by anyone who wants to say something about aviation safety without revealing his or her identity. This anonymity is important; without it the perpetrator of a near mishap might keep silent. First-person-singular "confessions" tend to be candid descriptions of human error which, when passed on to others, are extremely valuable training tools. Learning from the mistakes or near mistakes of others is beneficial. Anymouse reports can be submitted to *Approach* magazine, which features a section on them in each issue, to squadron/unit safety officers, or to the CO directly.

GRAMPAW PETTIBONE

Published since 1917, now under the cognizance of the DCNO (air warfare) and the commander, Naval Air Systems Command, *Naval Aviation News* is the oldest and perhaps most prestigious naval publication. It is not a safety publication per se, but it does feature the most widely recognized character in naval aviation safety: Grampaw Pettibone.

In each issue several mishaps are described in narrative form in the "Gramps" section. The stories are accompanied by illustrations from the talented hand of Robert Osborn, who helped create

Flying is not inherently dangerous, but it is mercilessly unforgiving of human error. (U.S. Navy)

"Gramps" during World War II and who still draws him. Critical remarks in Gramps' inimitable, acerbic style follow the narrative. This popular and widely heralded "column" is as enlightening as it is entertaining. Grampaw Pettibone is a great naval tradition.

A flyer hears many safety axioms throughout his or her career. Regardless of the programs of a particular command, one axiom, sooner or later, becomes firmly embedded in the aviator's mind: "Flying in itself is not dangerous, but it is mercilessly unforgiving of human error." Good words to remember.

NATOPS AND SAFETY

"Naval air training and operating procedures standardization program" is a stilted phrase, easier to manage in acronym form: NATOPS. This staple program of naval aviation has far-reaching, long-lasting connotations for each and every naval aviator. NATOPS sets down, lucidly and intelligently, step-by-step procedures for operating aircraft and accomplishing other aeronautical functions.

The importance of NATOPS may not reach Biblical dimensions, but a thorough if not commanding knowledge of NATOPS procedures is essential for the flyer who wants to become a genuine professional.

Two facts give quick proof of the value of NATOPS:
—Before the navy established the NATOPS program in 1961, the aviation mishap rate was nearly twenty accidents per one hundred thousand flight hours.
—In 1983 that rate had dipped to approximately four accidents per one hundred thousand hours

In large part, NATOPS is credited for this substantial reduction in loss of material assets, not to mention the savings in terms of lives.

MANUALS

Instrumental to the program is the NATOPS flight manual. One flight manual is published for each model of aircraft, and every flight manual contains standardized ground and flight operating procedures, training requirements, and technical data. The information in the flight manual is necessary for the safe and effective operation of the aircraft. Flight manuals also have indispensable pocket checklists for use in the cockpit.

A NATOPS manual, differentiated from a NATOPS flight manual, is issued for special operations that lend themselves to standardization, such as air refueling, carrier, instrument-flight, and LSO activities.

KEY NATOPS GROUPS AND INDIVIDUALS

Natops Advisory Group

This is comprised of representatives from the staffs of the CNO, commandant of the marine corps, and other major commands subordinate to these two. It monitors NATOPS to see that it functions properly. The Naval Safety Center, which is an advisory group member, is responsible for apprising other members of the effectiveness of the program.

Cognizant Command

This is an advisory group member designated by the CNO and responsible for specific portions of the program.

CNO NATOPS Coordinator

This officer is an experienced aviator tasked with overall NATOPS duties. While he monitors virtually all phases of the NATOPS program for the CNO, he also works in conjunction with the navy tactical support activity (NAVTACSUPPACT). Selected NAVTACSUPPACT officers act as CNO NATOPS coordinators and actively participate in NATOPS review conferences. NAVTACSUPPACT produces NATOPS manuals, coordinates the production of NATOPS flight manuals with various commands and contractors, and monitors the production, printing, and distribution of all NATOPS publications. It also promulgates a monthly report on cognizant-command and model-manager assignments.

NATOPS Coordinator

Like his CNO-level counterpart, this officer is a veteran flyer. Assigned to advisory-group member staffs, he coordinates the command's NATOPS program and maintains liaison with other coordinators as well as with the navy tactical doctrine activity. The coordinator sees to it that an annual review is made of each NATOPS evaluator's activities.

Model Manager

This is a unit designated by a cognizant command to administer the NATOPS program for a specific aircraft model. The model manager compiles proposed NATOPS changes and recommends the convening of review conferences when appropriate. Model managers review publications for accuracy and to ensure that they are in accordance with the latest approved operating procedures. Model managers coordinate the efforts of other NATOPS evaluators working on the same aircraft model to achieve standardization. The NATOPS evaluator for an aircraft model within the cognizant command is attached to the model-manager unit of that model.

Evaluator Unit

This is a squadron, unit, or air station that has been assigned responsibilities for NATOPS matters by the advisory group member.

NATOPS Evaluator

This person is a pilot or crewmember, highly qualified in a specific model of aircraft, occupying a primary billet in a unit designated by an advisory group member. Every year he evaluates NATOPS instructors in squadrons or units of the same model within the same major command. The evaluator checks flight crews at random in each instructor's unit to measure overall adherence to NATOPS procedures. He is also responsible for continuously reviewing manuals and certain maintenance and warfare publications to achieve standardization.

NATOPS Instructors

A NATOPS instructor evaluates all flight crewmembers within his squadron or unit and keeps the CO informed of all NATOPS matters.

NATOPS—EVERYONE'S BUSINESS

All members of the naval aviation community are encouraged to help update the NATOPS program. Procedural flaws or conflicts can be reported anonymously if an individual desires. OPNAV Instruction 3510.9 is the guiding directive for the program. It states, "NATOPS publications must have inputs from many sources in order to maintain the effectiveness of the program. To accomplish this, anyone in the naval establishment who notes a deficiency or an error is obliged to submit a change recommendation. The participation of the individual in this program of continual manual improvement is imperative."

Steps for initiating changes are outlined in the instruction. There are two ways to initiate a change to a specific NATOPS manual or flight manual. Routine changes are submitted to the model manager for consideration at a NATOPS review conference. Urgent changes are submitted through the cognizant command and approved by the CNO NATOPS coordinator for quick distribution to the fleet.

The essential purpose of the NATOPS program has not changed since it inception. It was created to provide the safest and best aircraft training and operating procedures. It has proven to be an all-encompassing management tool that continues to enhance the operational readiness of our naval aviation forces.

10

Aerospace Medicine

Revised by Captain John B. Noll, MC, USN

Aerospace medicine is a specialized field dedicated to the safe adaptation of men and women to flight, be they aviators or astronauts. Once airborne, an individual is subjected to the stresses of acceleration, deceleration, thermal extremes, radiation, vibration, weightlessness, and confinement. The ultimate goal of the aerospace medical specialist is to integrate clinical and preventive medical requirements so that the aviator/astronaut can perform his or her mission effectively and safely.

FLIGHT SURGEONS

The question is often asked, "Who do flight surgeons operate on?" The answer is, "Usually no one." The use of the term *surgeon*, especially in a military context, goes back well over one hundred years, when all military doctors were called surgeons. The word *surgeon* was also used as part of a rank. Considering that a surgeon's main task was amputations, the term was not exactly inappropriate. Until World War II all naval physicians were com-

Aerospace Medicine (U.S. Navy)

missioned as surgeons, assistant surgeons, or acting assistant surgeons.

There have been some interesting events in the evolution of our present-day aerospace-medical facilities. The navy's interest in aviation medicine dates back to 29 April 1922, when five naval medical officers graduated from the U.S. Army School of Aviation Medicine at Mitchell Field, Long Island. In January 1927 the navy began training its own flight surgeons at the Naval Medical School in Bethesda, Maryland. But in 1935 naval medical officers were again sent to the Army School of Aviation Medicine for their training. This continued until November 1939, at which time the navy opened the School of Aviation Medicine at the naval air station in Pensacola. Over 170 naval medical officers had received training in this specialized branch of military medicine by the start of World War II, and additional numbers were rapidly trained to meet the growing need for their services. Their great value was proven during the war.

Today the young medical officer who wants to become a naval flight surgeon must first volunteer for such duty and then complete at least one year of graduate medical education. The candidate, like his line-officer counterpart, must then be found able and physically qualified to fly. The prospective flight surgeon must pass the same flight physical standards that are applied to all aviators, except for modified requirements for visual acuity. Failure to meet the visual acuity standards does not disqualify him from flight training but may preclude solo flights.

Once chosen, the student is ordered to the Naval Aerospace Medical Institute in Pensacola, where he receives six months of training as described in chapter 2. The designated naval flight surgeons are then assigned to a variety of navy and marine corps air activities. These include air stations, major air command staffs, carriers, large aviation commands such as air groups and wings, and some specialized squadrons such as the Blue Angels.

The physician attracted to aviation medicine is not commonly one who seeks a career in that specialty. More often than not, he chooses a flight-surgeon tour when he is sifting options or eager for adventure, travel, and interaction with exciting people in an

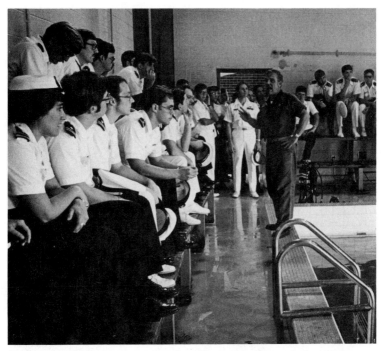

Prospective flight surgeons attend a class on water survival at Pensacola. (U.S. Navy)

exciting business. At the end of the first tour, the flight surgeon is offered additional training in the clinical area of his choice, an option exercised by the majority. A smaller percentage continue in aviation medicine. A few will select it for a career. The total number of flight surgeons will vary with the needs of the service, but the community is usually around three hundred strong.

The duties of both the naval flight surgeon and the dual designator* are augmented by those of qualified medical service corps officers (physiologist/experimental psychologist) who have also

*A dual designator is an officer who is qualified as both a naval aviator and a flight surgeon. Dual designators are often assigned as research pilots.

completed the syllabus offered at the Naval Aerospace Medical Institute.

THE FLIGHT SURGEON'S ROLE

The flight surgeon's job is multidimensional. His primary role is to be totally involved with the primary goal of the aviation community—takeoff, completion of an assigned task, and safe return to ship or base. This requires more than just healthy, qualified aviators. Aircrews and ground personnel must be just as qualified, motivated, and healthy as pilots. To accomplish the primary goal, all aviation personnel should be knowledgeable in those areas where

Two flight surgeons return from a flight in an F-14 Tomcat. One is a dual designator who is both a qualified pilot and a flight surgeon. (U.S. Navy)

psychological, medical, and/or physiological phenomena affect flight performance and mission accomplishment. The flight surgeon is the natural interface between the practice of medicine, the science of safety, and the business of naval aviation.

Accident prevention and membership in AMBs and field flight evaluation boards/field naval aviator evaluation boards (FFEBs/FNAEBs) are collateral tasks. As one who is qualified to analyze the human body, the naval flight surgeon is a very important link in the navy's safety program.

To accomplish these tasks and to be effective the flight surgeon must work, play, and fly with his people. He must know his personnel and their problems for the overall understanding of a situation and to be able to detect departures from normal patterns. As one CO put it, "If one of them has a headache, the flight surgeon should know it first!"

RELATIONS WITH THE FLIGHT SURGEON

One of the special associations pilots and NFOs have is their close relationship with their flight surgeon. Unfortunately many young aviators—and a few old ones—mistakenly see the flight surgeon as a policeman, the "company doc." No doubt this erroneous impression is in part fostered by the fact that flight surgeons have to decide who is physically qualified to enter aviation training. Once in naval aviation, the flight surgeon's job is to keep flight personnel alive, safe, and flying. He is their personal consultant on all medical and aeromedical problems.

The flight surgeon is also available to discuss a variety of topics, whether it is a child's immunization schedule or how to combat fatigue on long-range flights. Aviation personnel should take advantage of the flight surgeon's presence—get to know him and allow him to get to know them.

KEEPING FIT—A MENTAL AND PHYSICAL JOB

With each new aircraft or mission an aviator is assigned to, his physical and mental tasks are more complex and demanding. His environment will always be hazardous. The need for a sound physical-fitness program coupled with a healthy mental attitude is obvious to all who are knowledgeable about naval aviation.

A mature individual is aware of his strengths and weaknesses. He will develop programs to enhance his strengths and work on his weaknesses. So it is for the professional pilot and NFO.

It is assumed that such officers already have motivation, a competitive spirit, a high IQ, and physical ability. But, being human, they are liable to weakness.

ATTITUDE TOWARD HAZARDOUS DUTY

Flying may appear glamorous, but it is often hazardous. Landing an aircraft on a rolling, pitching carrier deck on a moonless night calls for the highest order of airmanship, alertness, and courage.

The pilot must perform his duties with a healthy respect for their inherent hazards. If hazards dominate his thoughts, a man's faith in himself and his aircraft will suffer and his airmanship will be adversely affected. A proper perspective is beneficial; it discourages dangerous maneuvers and flight operations that demand too much of the pilot or the aircraft.

One keeps hazards at a minimum by continually adhering to safe rules of flight, by demanding sound maintenance procedures, and by respecting aircraft and individual limitations. Statistics have proven that all but a small percentage of naval aircraft accidents can be prevented.

FEAR

Fear is an emotion that most men experience, and aviators are no exception. In fact, the so-called fearless student aviator is a potentially dangerous pilot, for he may be tempted to exceed the limits of his aircraft or his abilities. There are two kinds of fear—the immediate fear that comes when an emergency arises, or the subconscious fear, whose effects on a person are more subtle.

The first kind of fear may be beneficial. It is the body's way of summoning together all of its resources to meet an emergency. It can make an individual sharp and enhance his alertness at the controls. It can make him work harder and longer. On the other hand, this kind of fear may cause vertigo, nausea, or blurred vision.

It may seize the mind and cut the individual off from his environment. He may freeze and become incapacitated. The aviator must recognize these symptoms and take control. He must make fear work for him and not against him. Once the emergency has passed he must recognize what has occurred and learn from it. To deny that he is capable of fear is a liability he can ill afford in the profession he has chosen.

The second type of fear mentioned above may become chronic, expressing itself indirectly in various neurotic symptoms and attitudes. Unlike the cause of immediate fear, the cause of subconscious fear is usually not apparent. Its symptoms are more prolonged; they follow a person around, go home with him, make his life a misery. Sometimes they drive him to have the very accident he fears. Some of the more common symptoms of subconscious fear are loss of appetite, bad dreams, and an almost constant depression. A flight surgeon can help the individual deal with such problems by talking with him. With improved insight, the individual's fears may cease to exist. But if he ignores such problems, he may jeopardize himself, his family, and his squadron mates.

FLYING FITNESS—PERSONAL CHOICES

To fly today's high-performance aircraft and stay alive, flight crews must be as fit as their aircraft. The common cold, indigestion, and other minor illnesses that do not seriously impair individuals in other pursuits may be incapacitating to flying personnel. Inadequate nourishment, lack of sleep, excesses that lower efficiency, and preoccupation with extraneous matters are incompatible with flight safety and mission accomplishment.

Fatigue, Rest, and Sleep

Fatigue is a significant problem in naval aviation and all too frequently a contributing factor in mishaps. Three main causes of fatigue are generally recognized: stress, anxiety, and fear. There are three types of stress:

—Physical stress, caused by poor lighting, heat, noise, vibration
—Physiological stress, caused by hypoxia, lack of sleep, strenuous work
—Emotional stress

To some degree stress, anxiety, and fear all contribute to fatigue. Certain types of flying, particularly stressful, result in fatigue even though they require little physical activity. Formation flying, night flying, and flying in bad weather are a few examples of stressful situations.

Some of the more significant effects of fatigue are lowered physical efficiency, reduced night vision, decreased "G" tolerance, increased susceptibility to vertigo, and an unconscious lowering of performance standards. Many of these effects feed upon themselves. Fortunately there is evidence showing that a fatigued but sufficiently motivated individual has the ability to mobilize his skills for brief critical periods, as during landing.

Fatigue can either be temporary or chronic. Unlike temporary fatigue, the chronic type is not relieved by a good night's sleep. The individual suffering from chronic fatigue may have any or all of the following symptoms: increased tension, irritability, reduced aggressiveness, loss of appetite, lowered morale, and insomnia. If an aviator also has abnormal fears and trembling, he is likely to be suffering from combat fatigue—a form of extreme chronic fatigue.

There are several things an aviator can do to increase his resistance to the effects of fatigue. He must recognize the existence and ramifications of fatigue, and he is wise to maintain good physical fitness through proper sleep, diet, exercise, and recreation. Use of the required equipment and strict adherence to NATOPS procedures are a must. The wearing of properly fitted G-suits by fighter/attack pilots and strict adherence to the rules of oxygen usage also reduce fatigue. On long flights, when it is not possible to move about frequently, stretching the neck, back, and leg muscles can fight fatigue.

Adequate sleep and rest restore the body's mental and physical energy. While "cat naps" can help eliminate temporary fatigue, approximately six to eight hours of uninterrupted sleep a night are needed to prevent the onset of chronic fatigue. Flight schedules should accommodate this need. No flight personnel should be scheduled for continuous-alert and/or flight duty for more than eighteen hours. A crew rest period of fifteen hours minimum should be provided following such duties.

Exercise

There is no question that a planned exercise program contributes to one's physical well being by relieving nervous tension and mental fatigue and improving morale. The flight surgeon can give advice on exercise programs geared to a person's needs and age. Experience has shown that spurts of exercise may be harmful and are less effective than a planned, regularly scheduled program. For those engaged in highly competitive and tiring sports, adequate rest (at least twelve hours) is essential before flying.

Physical Achievement Requirements

Flight crewmen should be aware that all naval personnel must maintain certain physical standards. (The SECNAV Instruction 6100.1 series and NMPC Instruction 6100.2 series give details.)

Diet

An aviator should eat a balanced diet that fulfills his daily caloric and nutritional needs. Two or three meals per day should be eaten in relaxed and pleasant surroundings. Fliers should not attempt to lose weight by skipping meals but rather by reducing their caloric intake at every meal. Skipping meals merely reduces alertness and therefore increases the hazards of flying. Time and money should not be wasted on diet pills, sweat belts, and vitamins. Fads seldom work, and many have adverse physiological effects. Over the years a significant number of accidents have occurred partly because of missed meals or unwise diet practices.

A flight surgeon can help flight personnel set up a sensible meal program that will keep them flying safely and benefit them in the long run.

Flight Lunches and Food Poisoning

The flight or box lunches frequently used for in-flight meals are a potential source of food poisoning, which causes nausea, diarrhea, and vomiting. Food poisoning can readily disable an aircrew and result in mission abortion or worse. The most common form of poisoning, that caused by the toxins produced in food contaminated with the staphylococcus bacteria, commences about two to four hours after the food is eaten. However, symptoms may appear as early as one hour after eating. Food is usually contaminated during its handling or preparation. The contaminating organisms continue to grow, particularly when the food is not refrigerated. Food kept at a temperature above 40 degrees Fahrenheit should be eaten within three to four hours. To avoid food poisoning in flight, follow these simple rules:

—Do not consume food after the expiration date stamped on the box.
—Do not remove food from the aircraft and eat it after the flight.
—Store box lunches in a cool place in the aircraft—not near hot radio and other electronic equipment.
—Order two kinds of box lunches so the pilot can eat one and the copilot the other. This lessens the possibility that both will eat contaminated food.
—Separate the eating times of the pilot and copilot by at least one hour.

Medication

Taking medication without a prescription is foolhardy for anyone; for flying personnel it can be extremely hazardous. Aviators should even avoid taking over-the-counter drugs. After one accident, a pilot's personal effects were found to include muscle relaxant pills of a type known to reduce "G" tolerance significantly. After another fatal accident, it was found that the deceased pilot had been taking a common over-the-counter cold remedy. Although the AMB could not state that these accidents were actually caused by the medications, they were considered to be contributory factors. It is a fact that even such simple remedies as aspirin or cold tablets can impair the fine coordination and concentration required in flight. It is foolish to fly with a cold or a case of "GIs" (gastrointestinal disorder); it is irresponsible to fly with a minor illness under the influence of medication.

If a pilot or flight crewman must see a physician not trained in aviation medicine, he should report back to his flight surgeon prior to resuming flight. While most physicians are well trained and competent, few are aware of all the medical problems peculiar to the aviation environment.

In general, an aviator should not fly for twenty-four hours after taking any medication, unless the flight surgeon prescribing it has indicated otherwise. If he is sick, he shouldn't fly; if he flies, he shouldn't take medications without consulting the flight surgeon.

Alcohol

Alcohol is a potentially addicting drug. It affects memory, muscular skill, sensory acuity, and judgment. In addition, it can dehydrate the body (see Heat Injury, below). Alcohol therefore has no part in aviation, where the highest alertness, muscular coordination, and judgment are required. Personnel in a flying status whose faculties are impaired even to the slightest degree by alcohol or its after effects should be grounded until fully recovered. Alcohol is removed from the body at the rate of approximately one-third of an ounce per hour. It takes the body three hours to burn up the alcohol from a highball containing two ounces of 100-proof whiskey (one ounce of pure alcohol). At least twelve hours should elapse between the consumption of alcohol and the assumption of aircraft control.

Some aviators seem to think that a few beers are safe and that alcohol applies to only "hard liquor." This is not so. A pint of beer, a five-ounce glass of table wine, and a jigger of 86-proof whiskey contain the same amount of alcohol. In any case, moderation should be the rule during off-duty hours.

Illegal Drugs

There is no place in today's navy for those who use marijuana or other illegal drugs. They must never be taken by flying personnel. Those who have taken illegal drugs, either deliberately or inadvertently, must inform their flight surgeon.

Tobacco

Tobacco smoke contains a number of irritating substances, such as aldehydes, ammonia, and tobacco tars, which account for the burning effect it has upon the eyes. Smoke also irritates the membranes lining the mouth, upper air passages, and lungs. They may cause inflammatory reactions, which include the simple "smoker's hack," bronchitis, chronic thickening of the lining of the bronchial tubes, and most serious of all, emphysema. However, just the simple cough can prove to be a real problem to the aviator on 100 percent oxygen.

There is no question that cigarette smoking helps cause lung cancer as well as cancer of the lip, mouth, and larynx. Children of smoking parents are more prone to respiratory disease than those whose parents do not smoke. Personnel working in cold environments are more susceptible to hypothermia and frostbite if they smoke. Chewing tobacco clearly causes lip and mouth cancer.

Coffee/Caffeine

Coffee in moderate amounts is a pleasant and slightly stimulating beverage, because it contains caffeine, a white, crystalline substance that affects the muscles, nervous system, kidneys, and heart. One six-ounce cup of coffee contains about 100 to 150 mg of caffeine. An excess of caffeine can reduce muscle coordination and cause tremors, heart irregularities, irritability and, in some people, depression. These adverse effects are aggravated by excessive smoking. What is an excessive amount of coffee? This varies with the individual; ten cups a day is considered the maximum most people can tolerate without adverse effects. Remember that caffeine is also present in chocolate, some cola drinks, and tea.

Eye Hygiene

Good eyes and excellent vision are prerequisites for aviators. Here are some tips for eye protection:
  —Cheap sunglasses should be avoided.
  —Exposure to direct sunlight or overexposure on the beaches, ski slopes, etc., should be avoided.
  —Adequate lighting should be used for reading.
  —Protective gear should be used when using power tools, sanding, etc.
  —Aviators should never apply drops to their eyes without consulting their flight surgeon.
The radar operator should heed the following advice:

—Oxygen should be used above ten thousand feet to avoid oxygen deficiency.

—Extraneous light should be eliminated from the cockpit.

—Rapid change and marked contrast in intensity gain on the screen should be avoided.

—The radar eye hood should be used for proper eye distance and to increase eye comfort while viewing the screen.

Sunglasses

One of the most useful items an aviator has is a pair of sunglasses— in the air or on the ground. Even though sunglasses reduce the distance at which other aircraft can possibly be detected, this is compensated for by the reduction of glare, lessened eye strain, and generally increased comfort. Thus sunglasses may give an aviator an even greater detection distance during extended periods of flight.

To derive the largest benefit from sunglasses, the aviator should note the following points, many of which also apply to helmet visors:

—Sunglasses should be worn when flying in the glare at high altitude, above cloud cover, or on clear, sunny days. They should be removed when moving from bright illumination to dim light.

—Studies have shown that night vision decreases with age, especially after age forty, and that night vision can be impaired by as much as 30 to 50 percent if the eyes are exposed to high-level illumination during the day. Hence, personnel flying at night should wear sunglasses during the day if they are on beaches, ski slopes, etc.

—Sunglasses should not be worn when flying in or under heavy overcast conditions.

—Sunglasses should not be worn at twilight or at night and should never be put on to avoid a glare at night.

—Polaroid sunglasses should not be used when flying, since they only reduce glare when the head is in certain positions, and because they leave blind spots.

—Official flight sunglasses of the highest optical quality are available through the supply system. The flight surgeon can order prescription glasses if necessary.

Hypoxia and Hyperventilation

Hypoxia is defined as a lack of oxygen in the blood stream which if uncorrected leads to severe bodily disturbances. Hyperventilation, excessive respiration that leads to abnormal loss of carbon dioxide from the blood, may cause dizziness, "air hunger," and unconsciousness. The professional aviator should have some understanding of hypoxia, hyperventilation, and the effects of anxiety upon each.

During normal ground-level activities, the rate and depth of breathing are controlled by changes in the acidity of the blood. During physical activity, carbon dioxide and lactic acid are both produced in large quantities, causing a slight increase in blood acidity. The blood, as it passes through the brain, stimulates a special group of cells known as the "breathing center," which automatically increases the respiratory rate, permitting the body to exhale larger quantities of carbon dioxide and inhale more oxygen. This increase in breathing in response to muscular work is known as physiological hyperventilation and will not, under normal conditions, produce any unpleasant or abnormal sensations. As the breathing becomes heavier, blood vessels and heart muscle dilate, effectively increasing the amount of oxygen transported to the brain and heart.

The body's regulation of breathing under conditions of rest and physical activity operates effectively until a cabin altitude of approximately ten thousand feet is reached. Above this altitude, if the subject is relatively inactive, breathing automatically increases in response to a significant drop in the oxygen content of the blood. This less efficient mechanism of breathing control operates through reflex centers in the blood vessels of the neck and results also in a physiological hyperventilation, but at a time when the subject is not producing excessive carbon dioxide.

During normal flight conditions, the body tends to limit the breathing rate to avoid excessive loss of carbon dioxide. Pilots frequently exposed to normal flight conditions tend to become more resistant to small losses of carbon dioxide. If a person does not consciously or unconsciously interfere with his breathing, the adverse symptoms of hyperventilation should not occur. However, if the aviator suffers from anxiety, fear, excessive stress, or unusual resistance to oxygen flow, the excessive loss of carbon dioxide may cause the blood to become slightly alkaline and the blood vessels in the brain to constrict, reducing the amount of available oxygen. Dizziness, tingling in the hands and feet, visual distortion, and the inability to think clearly—the most common symptoms of hyperventilation—may be experienced, singly or in combination. Since these are also symptoms commonly noted during hypoxia, and to a certain degree produced by the same phenomenon—a reduction in the oxygen supply to the brain—a pilot may think he is suffering from hypoxia. Numerous studies and experiences have shown that the most common symptoms of hypoxia are visual distortion, dizziness, lack of muscular coordination, inability to think clearly, tingling in the arms and legs, apprehension, and in moderately advanced cases, characteristic bluish discoloration of the lips and fingernails.

Every pilot is aware of the insidious and potentially serious effects of hypoxia on his alertness. Much of the physiological training given to a pilot is designed to acquaint him with the symptoms of hypoxia and the measures to counteract it. But pilots do not receive equally intensive training on the effects of hyperventilation. Consequently, they are not always aware that this condition, while closely resembling hypoxia, is not as critical and may be fairly easily corrected. Hyperventilation occurs most often in the inexperienced or inadequately trained pilot. It is more frequently seen in students using oxygen systems for the first time, either in the air or in a low-pressure chamber during physiological training. It also occurs when a pilot is pressure breathing, and any time the body resists the movement of oxygen through the system. Regulated breathing is a remedy for hyperventilation.

Since hyperventilation is frequently precipitated by anxiety, it occurs frequently in a very tense and worried pilot who is under extreme emotional stress. He may be under stress when transferring to a new high-performance aircraft, using new equipment, or being exposed to a new or dangerous situation.

To complicate matters, various noxious agents and contaminants occasionally present in liquid oxygen systems or cockpit air may also cause an involuntary increase in breathing, either by stimulating the breathing center of the brain or by causing hypoxia.

To prevent or alleviate the symptoms of hyperventilation, an aviator must be able to recognize them (see table 10–1) or at least suspect their presence, and he must be well trained in breathing control. The symptoms of hyperventilation may easily be reproduced on ground or in a low-pressure chamber by simply taking four or five very deep breaths and noting the rapid onset of such effects as lightheadedness and tingling. (This procedure should not be carried too far, since it can produce unconsciousness and convulsions in a few extremely susceptible individuals.) The symptoms thus brought on should be contrasted to the symptoms of hypoxia experienced by all pilots in low-pressure-chamber indoctrination programs.

Altitude hypoxia rarely occurs below ten thousand feet and seldom results from exposure times of less than thirty minutes between ten thousand and fifteen thousand feet. At the higher altitudes symptoms of hypoxia may occur very rapidly; above forty thousand feet they may show up within twelve to fifteen seconds. Since altitude hypoxia is critical, corrective action can be taken by switching the oxygen regulator to the 100 percent and emergency settings and by taking a single deep breath and holding it for about ten seconds. At the same time, the aircraft should be brought to a lower and safer altitude. As soon thereafter as possible, a rapid but thorough oxygen-equipment check should be made. The breathing rate should be reduced to six or eight breaths a minute. If hypoxia or hyperventilation is the problem, improvement will be noted within fifteen to twenty seconds, although complete recovery may not ensue for several more seconds or minutes. If no improvement occurs, the oxygen supply may be contaminated, and the emergency oxygen system should be activated as soon as possible.

By this time, the pilot should come closer to knowing whether his difficulty is hypoxia, hyperventilation, or contamination. If hypoxia is suspected, 100 percent oxygen should be used with or without the emergency setting, depending on the cause. If hyperventilation is the problem, breathing control alone will suffice. If it is contamination, the emergency oxygen system should be used for the duration of the flight, which should be terminated as soon as possible.

Blood Donations

The naval aviator's physical fitness is directly dependent upon the oxygen-carrying capacity of his blood. For this reason, naval aviation personnel are grounded for four days after donating one unit of blood (500cc). Carrier aviators are prohibited from donating blood less than four weeks before they begin carrier operations. (See OPNAVINST 3710.7 series.)

VERTIGO

Vertigo means dizziness, spatial disorientation, or a "feeling" that an airplane is flying in one attitude when it is actually flying in another. Vertigo has been a flight hazard since the first days of aviation. Some 10 to 15 percent of all flight accidents are related to vertigo.

These are conditions in which vertigo can occur:
—Flying in IFR (instrument flight rules) conditions.
—Rotating the head when pulling G's or during a turn. (This causes a coriolis, or tumbling, effect.)
—Flying in marginal VFR (visual flight rules) conditions where the horizon is indistinct or absent. This is known as "flying in a milk bottle" or "paint bucket." (In this case, confusion results from conflicting cues and information transmitted through sight, inner ear, balance and receptors in the muscles and skin—what flight surgeons call the proprioceptive sense.
—Flying by IFR and VFR at the same time. (This should *never* be attempted.)
—Flying out of a set of flares into darkness after a night dive-bombing mission.

Table 10–1. Differentiation of Symptoms Occurring At Altitude

Condition	Common Symptoms	Cockpit Altitude	Exposure Time	Oxygen Condition	Corrective Action
Hypoxis	Visual disturbance Dizziness Lightheadedness Confused thinking Tingling Cyanosis	Rare below 10,000' Occasionally between 10–15,000'	Indefinite At least 30 minutes	Oxygen generally not used No oxygen used or significant leak in system	100% OXYGEN and EMERGENCY REGULATOR SETTING
	Apprehension Sense of well being Heart pounding	Common above 20,000'	12 to 60 secs	Leak in oxygen system or loss of mask after decompression	Descend to safer altitude
		Always above 45,000' without pressure suit	1 minute or less	With pressure breathing equipment only	
Hyperventilation	Dizziness Lightheadedness Tingling Visual disturbances Muscular incoordination Confused thinking Fatigue Numbness	Any altitude	Within seconds	Under any condition, but most likely when pressure breathing	*Breathing Control* If in doubt, take one deep breath of 100% oxygen, hold breath for 10 seconds, and breath more slowly

Cause	Symptoms	Altitude	Occurrence	System condition	Corrective action
Contamination of oxygen system and/or cockpit air	Depends on contaminant Unpleasant odor and/or irritation Increased breathing initially, later depressed Visual disturbances Lightheadedness Confused thinking	Any altitude	May occur at irregular intervals or constantly	*On 100% oxygen* Usually in oxygen system *Cabin Air Only* In cabin *On Dilution Oxygen* In oxygen system or cabin air	Switch to EMERGENCY OXYGEN SYSTEM and disconnect aircraft oxygen system ABORT FLIGHT
Anxiety	Uneasy sensation Tenseness Lightheadedness Dizziness Visual disturbances Fatigue Tremors	Any altitude	Constant or precipitated by unusual situation within seconds	Under any condition	Recognition of problem Breathing control

—Target fixation, self-hypnosis, or overeagerness and over-concentration on low-bombing runs, causing late pull-outs.

—Night air-to-air refueling.

—Moonless-night carrier landings.

—Wearing gas-mask goggles, which distorts depth perception in helo pilots.

Regardless of skill or experience, every pilot has vertigo from time to time. Good physical health and plenty of rest will go far to

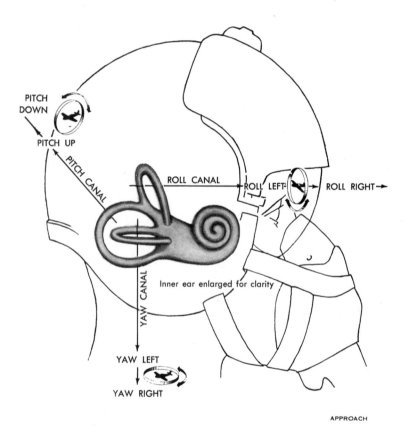

Fig. 10-1. Roll, Pitch, and Yaw Canals of the Inner Ear

combat it. Rapid head movements should be avoided when maneuvering, and the head should not be turned when pulling Gs. If a pilot gets vertigo, he must not panic but go to his instruments and believe what they tell him. Attention to scan is important, as is straight and level flight. Head movements should be minimized until things quiet down. As shown in figure 10-1, the movement of fluid in the semicircular canals of the ears can give false signals, and it takes time to return to equilibrium. If the pilot is flying in formation, the wingman should be made aware of the problem. Later, the incident can be discussed with the flight surgeon and squadron safety officer. Education about and recognition of vertigo offers the best insurance against accidents caused by vertigo.

All pilots should visit the navy's multistation spatial-disorientation demonstrator at the naval air station in Pensacola. There vertigo is induced, and pilots are given a taste of its effects.

HEAT INJURY AND THE AVIATOR

During the summer months or in tropical areas, aviation personnel run the risk of heat injury. The primary injuries of concern are as follows: heat cramps, which can be incapacitating; heat exhaustion, which can be incapacitating and may lead to heatstroke; and heatstroke, which can be fatal. Since 15 percent of heatstroke victims die, and since aviators who survive a heatstroke are permanently grounded, it is worth taking the time to understand heat injury.

The body defends against a rise in core (rectal) temperature by increasing the blood flow to the skin and by sweating. Environmental temperature, relative humidity, irradiation from the sun, and air movement determine the degree of heat you are exposed to. In an area of 100 percent humidity, where the temperature is higher than the body's and there are no air currents, the body cannot lose heat.

Any conditions that reduce the body's salt and water content make a person more susceptible to heat injury. Some common conditions are diarrhea, fever, recent immunizations, some medications that are not tolerant of heat, excessive alcohol consumption, and sunburn. Poor physical conditioning and lack of acclimatization also contribute to heat injury. Studies done during July in Pensacola have shown that pilots in full gear have lost up to 1.5 percent of their body weight on a routine sortie. If an aviator loses 3 percent of his weight, he is in danger.

Symptoms of heat injury, which may be present singly or in any combination, are as follows: cramps, lack of coordination, muscular weakness, poor judgment, mental dullness, irrational behavior, dizziness, throbbing headaches, and chilling sensations. If sweating stops, heatstroke has begun.

Here are some tips to help prevent heat injury:

—Use more salt on your food than normal, but do *not* use salt tablets.

—Wear clean clothing and a clean flight suit, since dirt can hinder the circulation of air.

—Compare your pre- and post-flight weights (stripped), remembering that a 3 percent loss is a danger signal.

—Cover the head and torso with light-colored material when exposed to the sun.

—If you experience any of the symptoms of heat injury, report to your flight surgeon.

—If your weekend included heavy exposure to the sun and alcohol, think twice about flying on Monday.

SCUBA POLICY

Scuba diving increases the amount of nitrogen dissolved in the tissues of the body. For aviators in flight, exposed to decreased barometric pressure, or in low-pressure chamber runs, there is an increased tendency for symptoms of "bends" (decompression sickness) to occur following diving. Therefore, naval aviation personnel should not fly within twenty-four hours following scuba diving or other compressed-air dives. Although a twenty-four-hour period may appear ultraconservative, it takes into consideration the general fatigue that normally accompanies such activities. And keep in mind the serious decompression sickness that several British divers sustained when they flew home at twelve thousand feet forty-eight hours after scuba diving. If circumstances dictate, an aviator may fly within twelve hours of scuba diving if he has been examined and cleared for flight by a naval flight surgeon. (See OPNAVINST 3710.7 series.)

ANNUAL PHYSICAL EXAMINATION

Whether a person is a candidate for flight training or an old hand with four thousand hours in the log book, he should look upon a flight physical not with apprehension but with confidence. Many people are anxious about the procedure, which causes their blood pressure and pulse to rise.

If a candidate is not able to pass the physical requirements, it is better that he know it and not risk his life later. The seasoned flier should keep in mind that a flight surgeon is committed to keeping as many aviators healthy and flying as possible. Experience has shown that most problems detected in a physical can be corrected or waived. It is better to spend some time resolving a minor health problem than to ignore it until it becomes major.

Aside from good general health, the two most important things an individual can do are rest before and relax during the examination. Don't go out on the town the night before and arrive tired and nervous. The key is to rest and relax.

Laboratory Tests

The aviation physical is a thorough general examination during which particular attention is given to the eyes, ears, and circulatory system. Blood is drawn for laboratory tests, which include the RPR (a test for syphilis). Urine is checked for any sign of urinary tract disease or kidney malfunction. Conditions affecting the whole body, such as diabetes, will frequently show up first in the urine. The chest X-ray is taken every three years until age forty-five, and yearly thereafter. A good X-ray will not only show symptoms of active or cured lung disease but will also indicate the condition of the heart, windpipe, upper spine, ribs, and diaphragm. Generally, the standard tests are all that are required, but others may be performed if there is reason.

Weigh-In

Next comes the weigh-in. The weight standards published by the Navy Medical Command are guidelines. Personnel in good physical condition who have large but not obese physiques usually pass. (The OPNAVINST 6110.1 series governs this.)

The ECG

Before the blood pressure and pulse, reclining and standing tests, aviators have a base-line electrocardiogram test. The ECG (or EKG) is repeated every three years until age thirty-five, after which it is done yearly.

Hearing Test

Because aviators are subjected to a great deal of high-frequency and high-intensity noise, they often begin to experience early loss of hearing in certain ranges. On the audiogram graph this shows up as the classic "aviator's notch." If protective measures are not taken, the problem can worsen.

Aviators working in noisy environments should have at least a yearly audiogram. If a person starts to lose hearing in the high frequencies (not in the spoken range yet), then by simply using extra protection and avoiding stereo headsets he can halt the damage.

Eye Examinations

Many of the optical measurements discussed below are obtained by the use of the armed forces vision tester, which has automated the examination process.

Visual acuity is the eye's ability to define an image. What the eye examination actually measures is the ability of the eye to bring an object into sharp focus on the retina. If the image that falls on the retina is blurred, the image "seen" by the brain is also blurred. The Snellen eye chart used in the examining room measures visual acuity.

Average visual acuity is 20/20. This means that an individual can see at twenty feet what most people can see at twenty feet. Vision 20/15 is better than average, 20/30 somewhat less than average. For entry into the flight training program 20/20 vision is required. However, service group 1 aviators whose vision has changed with age are allowed waivers that require the wearing of glasses while flying.

The eyes must line up in such a way that the brain sees only one, sharp image. If the eyes are not exactly aligned, diplopia or double vision results. No aviator can have diplopia, for obvious reasons.

The movements of each eye are affected by the actions of the six extraocular muscles (EOM) attached at intervals to the globe of the eye. These muscles are controlled by impulses from the brain, which always attempts to bring into focus one sharp image. To test for weaknesses or imbalances, therefore, the influence of the brain on these muscles must be eliminated. A device known as the phorometer provides different images to each eye—a line to one, a point of light to the other. Because there are two different images being "seen," the brain makes no attempt to superimpose them, no impulses are sent to the EOM, and the eyes assume a relaxed position, which may or may not be in alignment. If the muscles of each eye are perfectly balanced, the eyes will be aligned and the subject will see the line and the dot superimposed on one another. If some imbalance exists, which is usually the case, the line and the dot will be separated. To align them, prisms are used which bend the rays of light in such a way that the images are moved to the same point on each retina, thus superimposing the line and the dot. The amount of prism power needed to superimpose the images is then read from the phorometer and recorded. This gives the examiner a measure of whether the eyes tend to deviate in (esophoria), out (exophoria), up or down (right or left hyperphoria), and by how much.

Service group 1 aviators are allowed ten prism diopters (one diopter of prism power bends a beam of light one cm at one meter) of esophoria or ten diopters of exophoria as long as there is no double vision. Hyperphoria greater than 1.5 diopters is disqualifying.

The "prism divergence" test measures the power of the lateral muscles on the eye. The subject looks at a point of light while prisms of increasing power are placed in front of one eye; the prisms bend the light in such a way that the eye must rotate outward to maintain a single image in the brain. When the limit of outward rotation is reached, the result is double vision. At this point the amount of prism power is noted and recorded as prism divergence. In general, when the test is performed with the light thirteen inches from the subject, there must be at least twelve diopters of prism divergence to qualify for service group 1 standards. If esophoria exists, prism divergence at twenty feet is measured and must be equal to or greater than the measured esophoria. This test is performed only on the initial flight physical.

The ability of the EOM to rotate the eyes inward is measured by obtaining the point of convergence, that is, the nearest point directly in front of the eyes at which they can fix upon an object without diplopia. This is measured by advancing a small light down a ruler toward the nose and measuring the point at which the eyes can no longer maintain one image. This is the point of convergence. The interpupillary distance, or distance between the two eyes, is then measured. The point of convergence must be less than the interpupillary distance.

As stated before, all these tests are designed to detect eyes that, under any circumstances, may not be able to maintain a single image. The standards for flying personnel will vary with their jobs.

In the normal eye, rays of light originating at a point twenty feet or more from the eye are focused on the retina by the "resting" lens and cornea. However, to focus on an object less than twenty feet from the eye, the lens must add power. This process, called accommodation, is used by the aviator to adjust his eyes from a scan of the horizon to a scan of his instruments and is a measure of how long he can comfortably maintain a scan of these instruments. The testing procedure to measure accommodation no longer requires a ruler with small print on a sliding card. Instead, accommodation is now evaluated by the aviator's ability to read small print at a distance of sixteen inches. Student naval aviators must read the 20/20-size letters, while designated naval aviators must read 20/40-size letters, the standard-size print on navigation charts.

Aviators should not disregard the flight surgeon's advice to wear glasses. Glasses relieve the eyes of strain and allow them to function better in the cockpit. Fatigue or disease may result in lowered accommodative power and a decreased ability to alter scan rapidly or focus on close objects for prolonged periods of time; these objects, in the case of the aviator, are his instruments.

The need for perfect color and depth perception in the aviator is obvious. The tests are simple and direct. For color, the Farnsworth lantern is used, showing three cardinal colors, red, green, and white, in groups of two. All colors must be identified correctly. The Verhoeff stereopter consists of a box with a lighted background and three bars of different sizes placed in such a way that one is always either forward or back. There are four sets of positions for the stereopter, which is rotated up and down during the test so that, in all, sixteen positions are presented to the subject. The position of the bars in all sixteen must be properly identified to meet service group 1 standards. If eight out of eight are properly identified, the test is passed; if not, the sixteen are repeated and must be identified.

The peripheral fields of vision, so necessary to the fighter pilot, are outlined first by having the eye fix on a central object and then by bringing another object in from various angles until it enters the peripheral field. Not only the size but the shape of the peripheral field for each eye is then examined and interpreted. Any marked deviation from what is considered the normal size and shape of the field is cause for further investigation.

For persons thirty-five years old and older, an additional eye examination is done. The pressure in the eyeball is measured with an instrument called a tonometer, and the reading is known as the intraocular tension. An abnormally high pressure is seen in the disease known as glaucoma. Glaucoma, most common in older persons, leads to blindness unless it is detected early and treated properly.

The Final Tests

The remainder of the examination is essentially a simple, complete physical from head to foot. Do not lose sight of the fact that the flight surgeon is performing the physical in an effort to help you as well as the aviation community.

It is the responsibility of the individual aviator to arrange an appointment with the flight surgeon for his annual physical examination thirty days before or after his birthday. Should an officer fail to receive his examination within ninety days after his birthday, he must write a letter to the Naval Military Personnel Command

(NMPC) explaining the delay. Moreover, aviation personnel need to be aware that auditors periodically document the status of flight physicals. Those unfortunates who received flight pay for periods during which they were not administratively qualified (i.e., their physical was sixty days overdue) are forced to repay the money.

FLIGHT PHYSICAL STANDARDS

The Naval Medical Command prescribes the flight physical standards for each of the three service groups. When a pilot fails to meet the standards for his group, a report of medical examination standard form 88 is prepared by the flight surgeon and submitted to the command with the surgeon's recommendation. The command reviews the case and makes a recommendation to the NMPC. The recommendation may:

—Permit the pilot to continue in an unrestricted flight status if NMPC waives the standards.
—Restrict the pilot to flight duty in the next service group.
—Restrict the pilot to a light tempo of flight activity (this is done when a pilot is recuperating from illness or injury).
—Restrict the pilot to service group 3, requiring the presence of a service group 1 or group 2 pilot when flying (See NMPC Manual 1410100).

Naval aviators who fail to meet required physical standards are immediately suspended from further flight duties, and the circumstances are reported to NMPC (see NMPC Manual 1410100).

REVOKING FLIGHT STATUS FOR MEDICAL REASONS

When a physical or mental condition may be cause to revoke the flying authority and change the designator of an aviator, the chief of NMPC will usually request the nearest air command to appoint a board consisting of three flight surgeons to review the facts, examine the individual, and make a recommendation from a medical standpoint. If three flight surgeons are not available, the board may consist of other medical officers as long as at least one is a flight surgeon.

The board then submits its report to the chief of NMPC via the Naval Medical Command and the Naval Aerospace Medical Institute for disposition. When it appears that special studies or a more in-depth look is required, the Naval Aerospace Medical Institute will recommend that the aviator in question be ordered to appear before the Special Board of Flight Surgeons, located at the Naval Aerospace Medical Institute in Pensacola.

When an aviator requests appeal of a recommendation or decision of the board, the chief of NMPC may convene a formal board of senior flight surgeons at the Naval Medical Command in Washington, D.C. The decision of this board is final.

In cases where recommendations are made to terminate flight status, the chief of NMPC determines if the individual should be retained in the aeronautical organization or assigned outside the aeronautical organization.

11

The Enlisted Role

By Journalist Senior Chief Kirby Harrison, USN

Some of the most important people with whom the aviator has direct and frequent contact are the enlisted personnel who maintain, service, and man naval aircraft. The plane captain of an F-14 Tomcat fighter, responsible for ensuring that his aircraft is properly serviced and ready for flight, or the flight engineer who shares the cockpit with the pilots of a P-3 Orion are two obvious examples of such personnel. In addition to them, there are ordnancemen, electronics technicians, air-traffic controllers, and many others, all indispensible players on the naval aviation team. Still other enlisted men and women who are not technically part of the naval aviation rating structure are nonetheless involved in essential support activities.

Enlisted personnel make important contributions to the daily routine of navy flight operations. In an A-6 Intruder squadron, the efforts of no less than twenty-six enlisted personnel, each with differing skills, are required to keep each aircraft flying. A mini-

The flight engineer of a Lockheed P-3 Orion knows his aircraft inside and out. (U.S. Navy)

mum of eleven enlisted people are needed to get each aircraft off the ground. A squadron with ten aircraft may have as many as 250 enlisted personnel assigned.

Within the navy's aviation community, there are more than 140,000 enlisted men and women in aviation specific ratings. They range from the master chief petty officer to the nonrated airman apprentice recently out of basic training.

In aviation there are twenty-two specific ratings at the beginning level of third-class petty officer (E-4). Certain persons below the E-4 level may receive authorization as "strikers" in specific ratings based on formal classroom study and/or on-the-job training. Certain ratings change as the individual is promoted through the petty officer and chief petty officer levels. For example, enlisted persons going into the boatswain's mate (AB) rating will enter the third-class petty officer pay grade as an AB specializing in either catapult and arresting gear (ABE), fuels (ABF), or plane handling (ABH).

Upon promotion to the E-8 level, these individuals are considered to have had sufficient experience to take on a supervisory role in any or all of these areas. At this point, he or she will become an aviation boatswain's mate senior chief.

TECHNICAL TRAINING

Many rated men and women receive initial training at class A schools operated under the chief of naval technical training. They will first take one or both of the preparatory courses in aviation fundamentals and basic electricity and electronics. The former provides an introduction to naval aviation, while the latter imparts basic technical knowledge and skills in electricity and electronics.

Graduates of one of these schools or both then move on to class A schools, which teach more specific initial skills in the various ratings. Twenty of these schools are currently in operation at the NATTCs in Memphis and Lakehurst, New Jersey, and at the naval technical training center (NTTC) in Meridian, Mississippi. An-

The aviation boatswain's mate keeps things moving on the flight deck. (U.S. Navy)

other is a joint school located at Chanute Air Force Base in Rantoul, Illinois.

If a graduate of a class A school is slated for assignment to an operational squadron, he may be ordered to an FRS for further

Class A Schools	Location
Aviation maintenance administrationman (AZ)	Meridian
Aviation storekeeper (AK)	Meridian
Photographer's mate (PH)	Pensacola
Aerographer's mate (AG)	Chanute AFB
Aviation boatswain's mate, catapult and arresting gear (ABE)	Lakehurst
Aviation boatswain's mate, fuels (ABF)	Lakehurst
Aviation boatswain's mate, handling (ABH)	Lakehurst
Aircrew survival equipmentman (PR)	Lakehurst
Aviation ordnanceman (AO)	Memphis
Aviation structural mechanic, structures (AMS)	Memphis
Aviation structural mechanic, hydraulics (AMH)	Memphis
Aviation structural mechanic, safety equipment (AME)	Memphis
Aviation support equipment technician, electrical (ASE)	Memphis
Aviation support equipment technician, mechanical (ASM)	Memphis
Aviation antisubmarine-warfare operation (AW)	Memphis
Aviation electrician's mate (AE)	Memphis
Air traffic controller (AC)	Memphis
Aviation antisubmarine-warfare technician (AX)	Memphis
Aviation electronics technician (AT)	Memphis
Aviation fire-control technician (AQ)	Memphis
Aviation machinist's mate (AD)	Memphis

training. If not, he will be assigned directly to his permanent duty station.

Enlisted personnel may return to the technical training command one or more times during their careers for advanced and specialized training in class C schools or for short functional courses at class F schools. Further on-site training, which helps personnel deal with problems relating to specific weapons systems in current fleet use, is given in forty-six naval-air-maintenance group detachments operating worldwide.

ENLISTED CLASSIFICATION CODES

Within the aviation rating structure there are numerous subspecialties that are organized in a system of over nine hundred navy enlisted classification codes (NECs). These codes identify individuals with specialized expertise within individual ratings. Primary

Student air controllers practice the skills of their profession at the Naval Air Technical Training Center, Memphis, Tennessee. (U.S. Navy)

and secondary codes and corresponding job descriptions can be found in the Manual of Enlisted and Personnel Classification and Occupational Standards (NAVPERS 18068D). In the aviation-electronics-technician rating alone there are more than seventy separate NECs. Old codes may be dropped and new ones added in response to changing technology and the needs of the navy.

AIRCREWMEN

Individuals holding the aircrewman NEC are regularly assigned duties aboard both fixed- and rotary-wing aircraft. Missions include long-range patrol, antisubmarine warfare, airborne early warning, carrier onboard delivery, logistics/tactical support, and search and rescue.

Those who pass the rigid physical requirements for this duty attend the five-week naval aircrew candidate school at Pensacola. Much of the syllabus is similar to that used for AOCs. There is

These P-3 Orion crewmen are key players in the serious game of antisubmarine warfare. (U.S. Navy)

Aircrewman

considerable emphasis on physical training, land and sea survival, first aid, and aircraft familiarization.

After assignment to a squadron, the individual must qualify in the specific type of aircraft in which he or she will fly within eighteen months of reporting. Upon successful completion of all requirements, the appropriate 8200-series NEC is granted and the individual is entitled to wear aircrew wings.

Aircrew personnel represent a cross section of most of the enlisted aviation occupational fields. The only rating for which an aircrew NEC is mandatory is that of aviation antisubmarine-warfare operator.

Qualified enlisted personnel with good performance records may be selected to attend the Naval Air Crewman Candidate School at Pensacola. (U.S. Navy)

Aviation Warfare Specialist

AVIATION WARFARE SPECIALISTS

In 1980 the navy established a program with an appropriate wing emblem for those enlisted persons qualifying as aviation warfare specialists. Those wishing to win their wings in this program normally come from the group-nine (aviation) ratings. General requirements include an overall performance rating of 3.6 or better, nomination by the individual's CO, and a minimum of twenty-four months of sea duty (for rotational purposes) in an aviation assignment. Candidates, including those with nonaviation ratings, must be currently serving in deployable aviation billets, aviation billets within the shore establishment that support the air-warfare mission directly or indirectly, aviation billets allocated to vessels from which aircraft are operated or controlled, and billets nominated by COs and approved by the DCNO (air warfare) that contribute directly to the air warfare mission.

Specific requirements and conditions under which waivers may be granted are contained in OPNAVINST 1412.5.

The administration of training and testing is carried out at the local command level, and standards are kept high to maintain the integrity of the specialty. Individual qualification requirements may be approved only by those who have themselves previously qualified. Final approval is generally granted by a selection board at the local command level.

Aviation personnel who qualify are authorized to wear aviation-warfare-specialist wings, and the letters *AW* in parentheses are added to their rating designations. The specialty is not open to officers, but limited-duty officers and warrant officers who qualified prior to commissioning retain the privilege of wearing their wings.

Basic Parachutist

Parachutist

PARACHUTISTS

In years past, parachute riggers, now called aviation survival equipmentmen, were required to make at least one jump with chutes they had packed. This is no longer a requirement. The navy does, however, continue to operate a parachutist's program at the naval air station at Lakehurst, New Jersey, for those assigned to jump billets requiring this specialized training. The courses are attended by both officers and enlisted personnel, many of whom are not connected with the aviation community.

The basic naval-parachutist course (NP-1) runs for three weeks and includes five static-line jumps. Upon completion of this course, the student is qualified as a basic parachutist and may wear the silver wings signifying this accomplishment.

A basic parachutist may upgrade his qualification to parachutist by making five additional jumps for a total of ten.

The advanced naval-parachutist course (NP-2) involves further instruction in parachuting, and upon completion the student is certified for free fall. The prerequisite for this course is twenty-five jumps.

The jumpmaster course (NP-3) is the most advanced and results in certification as a jumpmaster.

RESCUE SWIMMERS

The rescue-swimmer program is designed to maintain a search-and-rescue capability for U.S. naval forces. Qualified volunteers who have completed aircrew training are assigned to the rescue swimmer program for one month. The training for this demanding work includes a rigorous physical regimen, seventy-five hours of pool work, extra first-aid instruction, and familiarization with all types of harnesses and water-survival equipment. Following res-

Rescue Swimmer

cue-swimmer training, candidates are required to complete the one-week survival, evasion, resistance, and escape school. Rescue swimmers must requalify annually at their squadrons and once every three years at the naval air station in Jacksonville or North Island.

A fully qualified rescue swimmer may be called upon to jump from a low hovering helicopter during daylight hours or be lowered by hoist at night into seas as high as twenty to thirty feet. He must be capable of swimming to a survivor and performing a rescue under hazardous, life-threatening conditions. There are more than one thousand billets for helicopter crewmen that require rescue-swimmer training.

NONAVIATION RATINGS

The hospital corpsman rating is not an aviation-specific rating. Within the rating, however, two NECs are especially important to the aviation community. These are the aviation-medicine technician (8406) and the aerospace-physiology technician (8409). Other corpsmen are also involved in the care of aviation personnel and their dependents at naval-air-station dispensaries and facilities worldwide. Each of our carriers may have as many as forty corpsmen aboard when it puts to sea.

Along with hospital corpsmen, other nonaviation personnel serve with aviation units. There are yeomen and personnelmen, for example, in every squadron. A helicopter mine-counter measures squadron includes several boatswain's mates who handle the gear used in sweeping operations. Aircraft carriers and naval air stations incorporate the entire range of enlisted ratings.

Upon assignment to an aviation unit, the pilot or NFO will invariably have a job supervising enlisted personnel. Because officers in flying status spend so much time flying and in related activities, they rely heavily on senior enlisted men and women to see that the squadron's day-to-day business is conducted efficiently and effectively.

An officer's interest in and attitude toward enlisted personnel can have a positive effect and result in enthusiasm that is contagious. A genuine respect for subordinates and the importance of their contributions to the team effort is an essential element of effective leadership.

AVIATION RATINGS

Aviation Boatswain's Mate (AB–ABE, ABF, ABH)

The boatswain's mate is responsible for launching naval aircraft quickly and safely from ship or shore and for conducting preflight and postflight checks of launching and recovery equipment. ABEs specialize in catapult and arresting gear, ABFs in fuels, and ABHs in directing and handling aircraft on deck as well as operating such rescue gear as cranes. At the E-8 level, all these personnel become simply ABs, capable of supervising in any or all of these specialized areas.

Aviation
Boatswain

Air Traffic
Controller

Aviation
Machinist

Air Traffic Controller (AC)

He is responsible for the safe, orderly, and speedy movement of aircraft into and out of landing areas. Landing areas may be an airfield or aircraft carrier.

Aviation Machinist's Mate (AD)

He inspects, adjusts, tests, repairs, and overhauls aircraft engines. He may specialize in reciprocating engines or jet engines, although the navy no longer maintains a school for reciprocating-engine training.

Aviation Electrician's Mate (AE)

He is capable of handling a wide range of electrical and navigation equipment found in naval aircraft, including power generators, flight instrument systems, and pressure-indicating systems.

Aerographer's Mate (AG)

He is the navy's meteorological and oceanographic expert, trained in the science of meteorology and physical oceanography.

Aviation Storekeeper (AK)

He makes sure materials and equipment needed for naval aviation activities are available.

Aviation Structural Mechanic (AM–AME, AMH, AMS)

He installs, maintains, and repairs the metal structures of aircraft, moveable aircraft parts and their control systems, and aircraft-body surfaces. He also maintains and repairs utility systems and safety devices. AMEs specialize in safety equipment, AMHs in hydraulics, AMSs in aircraft structures.

Aviation
Electrician's Mate

Aerographer

Aviation
Storekeeper

Aviation Structural
Mechanic

Aviation
Ordnanceman

Aviation Fire
Control Technician

Aviation Ordnanceman (AO)

He is in charge of storage, servicing, inspecting, and handling all types of weapons and ammunition carried on navy aircraft.

Aviation Fire Control Technician (AQ)

He is the electronics specialist responsible for the upkeep of weapons control systems on navy aircraft.

The aviation ordnanceman must be able to perform his job safely and efficiently. (U.S. Navy)

Aviation Support
Equipment
Technician

Aviation Electronics
Technician

Aviation
Antisubmarine
Warfare Operator

Aviation Support Equipment Technician (AS–ASE, ASH, ASM)

He operates, maintains, repairs, and tests the ground equipment used in handling, servicing, and maintaining aircraft and aircraft equipment. ASEs specialize in electrical components, ASHs in hydraulics and structures, ASMs in mechanical functions.

Aviation Electronics Technician (AT)

He maintains the advanced-technology radio, radar, and electronics equipment carried on aircraft.

Aviation Antisubmarine Warfare Operator (AW)

He operates airborne electronic equipment used in detecting, locating, and tracking submarines.

Aviation Antisubmarine-Warfare Technician (AX)

He is responsible for keeping antisubmarine-warfare systems and equipment in good operating order.

Aviation Maintenance Administrationman (AZ)

He performs the clerical, administrative, and managerial duties necessary to keep aircraft maintenance activities running smoothly.

Photographer's Mate (PH)

He operates cameras in a variety of assignments, ranging from

Aviation Antisubmarine
Warfare Technician

Aviation Maintenance
Administrationman

A photographer's mate aboard a P-3 Orion photographs a surface contact during an ocean surveillance flight. (U.S. Navy)

news events to photographs used in mapmaking and reconnaissance.

Aircrew Survival Equipmentman (PR)

He keeps parachutes and other aviation-survival gear, including life rafts, life jackets, oxygen breathing apparatus, signaling equipment, and air-sea rescue equipment, in proper working condition.

Tradevman (TD)

He installs, operates, maintains, and repairs training aids and training devices. (This rating is scheduled for elimination by 1988.)

PH

Photographer's mate   Aircrew Survival
Equipmentman   Tradevman

12

The Naval Air Systems Command

By William J. Armstrong

The Naval Air Systems Command, often referred to as NAVAIR, is responsible for developing, procuring, and supporting all the aviation systems and their related equipment used by the navy and marine corps, including airframes, engines, electronics, and most air-launched weapons. The command also oversees the life cycle of each aviation system, from research and development through disposal. The spectrum of NAVAIR's work begins with naval aviation's technology base, from which the command develops the systems that one day become operational. NAVAIR acquires these systems for the navy from private industry, tests and evaluates them, and then furnishes them to their users. From that point on, NAVAIR maintains the systems—and modifies them as necessary—until they become obsolete. NAVAIR is unique among the commands of the three services; neither the air force nor the army has one organization that covers the entire life cycle of its aviation systems. To perform its vast job, NAVAIR is organized as a three-

The Naval Air Systems Command (U.S. Navy)

The headquarters of the Naval Air Systems Command is located in Virginia, not far from the Pentagon. (U.S. Navy)

star command with headquarters in Washington, D.C., and twenty-nine field activities in the United States and abroad. The command has a personnel complement of some forty-five thousand and an annual budget of nearly seventeen billion dollars.

NAVAIR's work is wholly with the material side of naval aviation. The command does not operate the systems that it furnishes and maintains; no navy systems command does. The operating forces are the province of the CNO. Each systems command is primarily concerned with the material requirements of its systems. These requirements are based on the existing and projected needs of the fleet and are formally established by OPNAV. When an OPNAV directive calls for aviation material, the requirement is prosecuted by NAVAIR. However, NAVAIR's activities are not determined solely by directions from higher headquarters. On the contrary, the command not only develops aviation material for the fleet, it is responsible as well for conducting technological research and recommending and determining the feasibility of future systems.

THE HISTORY BEHIND NAVAIR

From the time the navy began using air power, some organization in the Navy Department has been charged with developing the best technology and hardware for integrating aeronautics and the fleet. From 1911 to 1921 the navy administered aircraft responsibilities as though aircraft were small ships; authority for them was divided among various bureaus. The Bureau of Construction and Repair was responsible for airframes; the Bureau of Steam Engineering for aircraft engines; and the Bureau of Ordnance for airborne weaponry. In 1921, when this system was acknowledged as unsatisfactory, the Bureau of Aeronautics was created. Into this bureau were drawn virtually all elements of naval aviation, and the organization had a direct voice in operations. The bureau's plans division established naval aviation's requirements; the material division was charged with meeting them. The only exception was ordnance: The bureau established aviation's requirements for weapons, but the Bureau of Ordnance continued to be responsible for ordnance. This arrangement prevailed until 1943, when four of the divisions of the Bureau of Aeronautics—plans, flight, training, and personnel—were transferred to OPNAV. They were used to form what is today the office of DCNO (air warfare), or OP-05. This transfer of functions left the bureau with its material division; all responsibility over operations was vested in OPNAV. The Bureau of Aeronautics' responsibility for aviation material remained basically unchanged until 1959. In the meantime, however, a question arose between the Bureaus of Aeronautics and Ordnance over guided missiles. Because early guided missiles were primarily aircraft loaded with explosives, the Bureau of Aeronautics developed them and referred to them as pilotless aircraft. Nonetheless, the Bureau of Ordnance was responsible for developing and furnishing naval weapons. In 1959, in an effort to resolve the difficulties posed by divided cognizance over a system of growing importance, the two bureaus were merged to form the Bureau of Naval Weapons. This bureau was disestablished in 1966, and in its place were created three systems commands: air, ordnance, and electronics. After 1943 NAVAIR retained all the responsibilities of the old Bureau of Aeronautics, plus responsibility for practically all air-launched weapons, including rockets and guided missiles.

ORGANIZATION OF NAVAIR

NAVAIR has a system of management called the program/functional matrix. The basic purpose of the program component of the matrix is to manage the coordination of technical and business aspects of acquiring and maintaining systems. It establishes central managers and their staffs, each charged with managing a specific program. Within NAVAIR, there are five programs: program management for aviation (PMA), air program coordination (APC), the Advanced Development Program Office (ADPO), system program or commodity area management (SPM), and weapon system management (WSM). None of these programs does by itself all that must be done regarding the system it manages. They all depend for technical support upon the NAVAIR groups that comprise the functional component of the command's program/functional matrix. These functional groups and their codes are:

—Assistant commander for contracts (AIR-02)
—Assistant commander for research and technology (AIR-03)
—Assistant commander for logistics/fleet support (AIR-04)
—Assistant commander for systems and engineering (AIR-05)
—Assistant commander for test and evaluation (AIR-06)

Unlike the program offices, which manage only one system or commodity area, the functional groups support all the command's programs.

Of the methods used by NAVAIR to establish central program authority, the PMAs are the most visible. All major system-acquisition programs have PMA status. A program is designated a PMA because of its cost, urgency, complex technology, or need for central management. When a program is selected for PMA status it may be anywhere from the exploration to the deployment stage. The program manager is responsible for planning, control-

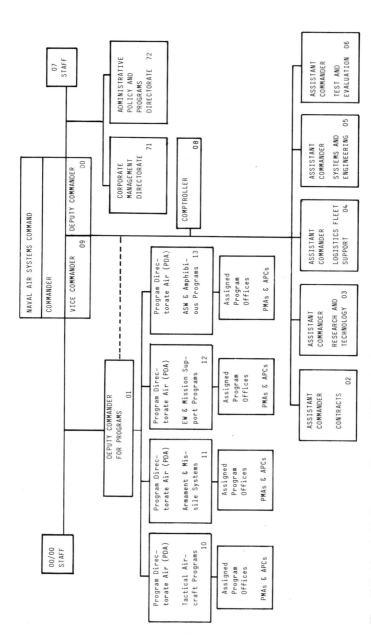

Fig. 12-1. Naval Air Systems Command Organization

ling, and directing his program. He does not, however, direct the purely technical aspects of his program; for technical support he must rely on the engineers and technical specialists in NAVAIR's functional groups. The program manager is, nonetheless, responsible for overall management of the technical, budgetary, and scheduling concerns of his program.

All PMAs are coordinated and supported by the deputy commander for programs (AIR-01). It is AIR-01's responsibility to provide command-wide coordination to ensure that uniformity is maintained among the policies and procedures of program management. Each PMA is assigned for coordination and support to an appropriate program directorate air (PDA), who is under AIR-01. There are four PDAs in NAVAIR: tactical aircraft programs (AIR-10), armament and missile systems (AIR-11), electronic-warfare and mission-support programs (AIR-12), and anti-submarine-warfare and amphibious programs (AIR-13). Ultimately, however, all PMAs are accountable to the commander of NAVAIR.

At any time, NAVAIR has several programs in PMA status. At the time of this writing there are twenty-three PMAs in NAVAIR.

When a PMA's objectives have been met and its system ceases to warrant intense management, the PMA is either disestablished or redesignated an APC, or its system is transferred to a WSM.

The APC is not used solely for redesignated PMAs. It is meant to emphasize and increase the visibility of certain programs that do not warrant PMA status. A program selected for APC status may be anywhere from the program initiation to the deployment stage. The authority and responsibilities of an APC are the same as those of a PMA; both are coordinated similarly by AIR-01. When the objectives of an APC have been met, its system may be transferred to either PMA or to WSM status; in either case, the APC is subsequently disestablished.

WSM is ordinarily employed to decentralize the management of systems no longer in production but still in use. Support can be furnished by one of NAVAIR's field activities. For example, the F-4 Phantom II is still in the fleet but not in production. Its WSM is located at the naval air rework facility at North Island. All WSMs are under the central control and coordination of the WSM project coordination office, which is APC-10 in NAVAIR headquarters.

At the opposite end of systems life cycles is the ADPO. The programs in ADPO status are usually in the research and technology stage. The ADPOs have command-wide support for their programs. They are managed by NAVAIR's AIR-03. When ADPO programs reach a point where acquisition and increased attention are warranted, they are transferred to the status of a PMA or APC. It should be noted, however, that very few of NAVAIR's designated programs begin as ADPOs.

The programs mentioned thus far are organized so that each deals mainly with a specific system. The system may be an aircraft, a missile, or a projected aircraft. In the case of system program management (SPM), however, we find an office directing what is called a commodity area. There are five of these areas in NAVAIR: crew systems (AIR-531), avionics systems (AIR-543), propulsion and power (AIR-536), armament systems (AIR-541), and weapons training (AIR-413). Their codes indicate that they belong to the functional groups for systems engineering (AIR-05) and logistics/fleet support (AIR-04). All but AIR-413 are commodity areas clearly related to engineering and are thus chartered within AIR-05. AIR-413 is a logistics function and therefore is chartered in AIR-04. The SPMs report to their respective assistant commanders. From their titles it is clear that the SPM commodity areas cover more than one major weapon system. Each area needs intense program management.

It is important to understand that only a small portion of the actual hardware NAVAIR develops and furnishes is built in-house. The same is true of many of the services the command performs. With only a few exceptions, the navy has procured most of its aviation material, and many services, from private industry. It follows that one of NAVAIR's major functions involves contractual agreements between the command and the private sector.

Because NAVAIR has so many contracts in force at any time and such a large number pending award, there is a functional group solely for contracts. The assistant commander for contracts (AIR-

02) coordinates contractual matters between private industry and NAVAIR offices. AIR-02 conducts all the negotiations and awards all of NAVAIR's contracts. He also represents NAVAIR in the development of procurement policy at the highest levels of government.

Within AIR-02 are the principal contracting officers (PCOs). They are the focal points for the program within the AIR-02 functional group. For example, there is a PCO for the F-14 Tomcat (AIR-214D), another for the Sparrow missile (AIR-21615). A PCO is a functional specialist whose responsibility it is to coordinate his specific program with the contract support of his group.

Various approaches to satisfying an operational requirement must be considered and the emerging designs tried and refined to ensure the correct choice of a weapon system. The choice and subsequent development, engineering, and testing of a weapons system is the responsibility of the assistant commander for systems engineering (AIR-05). The AIR-05 group has all the technological skills and knowledge necessary to develop and engineer a modern aviation weapon system for the navy. This group develops and publishes the specifications for the procurement and production of aircraft systems. All systems proposals are evaluated in AIR-05 to determine to what extent they meet the requirements for which they are proposed. In AIR-05 design, cost, and mission effectiveness are analyzed and evaluated. Each component of an aviation system—airframe, weaponry, countermeasures and support equipment, catapults, arresting gear, photography/reconnaissance equipment, and even meterological devices—is studied, engineered, and integrated into a system by AIR-05. The group also develops the plans for and conducts the engineering, development, technical evaluation, and production management of systems and equipment being acquired. AIR-05 informs OPNAV when operational evaluation should begin on aviation systems.

Once a system is acquired, AIR-05 provides engineering support throughout the remainder of the system's life cycle. The group establishes and manages the NAVAIR/Defense Department standardization programs and prepares all technical manuals for all aviation systems.

Within AIR-05, assistant program managers for systems and

engineering (APM,SE) liaise between the group and the command's program offices. Some of the most visible APM,SEs are assigned to AIR-511, the systems engineering management division, and are unofficially known as class desk officers. This term goes back to the days when the Bureau of Aeronautics was a purely functional organization without program offices. AIR-511 has an APM,SE for each aircraft system for which the command has a designated program manager. Other APM,SEs are found in the air-launched guided-weapons division of AIR-05 (AIR-542). An APM,SE is provided to each guided weapon for which NAVAIR has a designated program manager. APM,SE offices for antisubmarine warfare and electronic warfare are assigned within the avionics equipment division (AIR-549).

An assistant program manager is the coordinator of all technical support provided to his project by AIR-05. He is the leader of the program support team for his program. The several divisions in his group are responsible for giving technical support to NAVAIR programs through the appropriate APM,SE; administratively, however, the divisions are responsible to their superiors.

From the time that AIR-05 begins the hardware development of a weapon system, the logistics support for that system is also being developed. Preparation of an integrated-logistics-support plan for each aviation system and weapon and commodity is the responsibility of the assistant commander for logistics/fleet support (AIR-04). AIR-04 prepares plans in coordination with elements of NAVAIR and many command field activities.

Each integrated-logistics-support plan developed by AIR-04 discusses the maintenance, facilities, support equipment, parts, manpower, technical publications, and training needed for a system. Once a system or commodity is operational, any change proposed for it must be reviewed by AIR-04. If revisions are necessary, it is AIR-04's responsibility to make them. Finally, the group establishes policy for the disposal of surplus equipment in accordance with laws and regulations.

Within AIR-04 are assistant program managers for logistics (APM,L). Each APM,L is the leader of the integrated-logistics-support management team for his program. For example, there are APM,Ls within the airborne weapons logistics division (AIR-

420) for weapons, within the maintenance policy and planning division (AIR-411) for common avionics, and within the logistics management division (AIR-410) for aircraft systems. All logistics specialists in AIR-04 provide technical support to the APM,Ls, but they are administratively responsible to their superiors within their own divisions.

Engineering development and the logistics and maintenance support of aviation systems are only two of NAVAIR's responsibilities. No less important is the test and evaluation of systems that NAVAIR procures.

The assistant commander for test and evaluation (AIR-06) is the command's newest functional group. It was created in 1975 to consolidate the management of the facilities and resources of aviation test and evaluation. AIR-06 establishes the policies and procedures that govern the planning, conduct, and reporting of test and evaluation. Policy direction, management, planning, and functional support are provided by the group to the NAVAIR program/acquisition managers. These functions are also performed by AIR-06 for all navy test ranges and the navy's portion of the Department of Defense's Major Range and Test Facility Base. The group is also responsible for managing, directing, and administering the navy aerial- and surface-target program and the range-instrumentation system program.

In AIR-06's projects division (AIR-620) are assistant program managers for test and evaluation (APM,T&E). Like other program managers, the APM,T&Es provide technical support for the command's program/acquisition managers. AIR-06 certifies that all testing is completed and conducted in accordance with the test and evaluation master plan (TEMP). The group also makes sure test reports are available. Before a system can pass from technical evaluation to operations evaluation, AIR-06 must inform OPNAV of the system's eligibility for further testing.

The TEMP defines the testing to be done for the system being acquired. Plans for engineering and development, along with the technical content of the TEMP, are drawn by AIR-05; then the APM,T&E directs and manages the testing at one of the command's field activities or at a navy laboratory. Operations evaluation is done by the Operational Test and Evaluation Force after NAVAIR certifies to OPNAV that a system is ready.

Naval aviation continues to improve existing areas of technology and develop new areas. As mentioned earlier, one of NAVAIR's responsibilities is managing the technology base of naval aviation. The assistant commander for research and technology (AIR-03) manages the general areas of naval aviation's research and technology. The AIR-03 group does not develop systems. Its work is mainly with research, exploratory development, and advanced development of the concepts used by NAVAIR in the actual development of systems. AIR-03 writes the plans and programs necessary for research and development within the command.

The OPNAV requirements for advanced aviation systems are used by AIR-03 to determine its own objectives regarding advanced systems. One of these objectives is constantly to assess technological innovations and coordinate with higher headquarters to make sure they are reflected in navy planning. All suggestions for advanced air systems are approved or disapproved by AIR-03 through studies, the use of simulators and experimental prototypes, analysis, or feasibility demonstrations. When a program in AIR-03 is at the stage of development where it needs program management, an advanced-development program office may be established, but not all of AIR-03's work in the area of advanced development leads to such an office.

NAVAIR's highest authority is AIR-00. A vice admiral is principally assisted by a vice commander (AIR-09) and a deputy commander (AIR-07). The vice commander is ordinarily a rear admiral. The AIR-07 position, established in 1983, is filled by a senior civilian in NAVAIR—where the personnel ratio is around ten civilians to each military billet—who shares with the vice commander the responsibility for executive direction of the command as delegated by AIR-00.

Assisting the three commanders are various staff offices, the names of which are self-explanatory. Assigned directly to the commander are the Office of Counsel, the Small Business Office, the Office of Equal Employment Opportunity, and the Office of Patent Counsel. On the vice commander's staff are the military affairs officer, the naval-reserve air-systems-program officer, the career management officer for AED/AMD and aviation WO/LDO personnel, and the force master chief. Also on the AIR-09 staff are the safety officer and the inspector general. Under the direction of the deputy commander are the Civilian Personnel Program Office, the Legislative and Public Affairs Office, and the Readiness, Reliability, and Maintainability Office. The deputy commander's organization is itself divided into two directorates, corporate management and administrative policy and programs.

The comptroller (AIR-08) is senior financial adviser to the commander. He develops and implements policies and procedures for financial and resource management, and sees that the financial management of NAVAIR supports the command's mission and conforms with laws and regulations.

NAVAIR FIELD ACTIVITIES

Thus far only NAVAIR's headquarters has been discussed, but headquarters is hardly the sum of the command. NAVAIR has twenty-nine field activities that furnish technical support to the command's functional groups and program offices. This support covers the entire spectrum of weapon systems, from development to engineering, test and evaluation, and logistics. Field activities are decentralized components of NAVAIR. An immediate superior in command (ISIC) directs and controls the field activity. He is responsible for inspections and the planning of mission, facilities, and work load. He coordinates material and other elements of support. The primary support official (PSO) administers the field activity's funds, manpower, facilities, and material.

NAVAIR also receives technical support from navy laboratories and research and development (R&D) centers.

A brief description of NAVAIR field activities follows.

Naval Environmental Prediction Research Facility (NEPRF)

Located in Monterey, this facility conducts applied research in atmospheric science to improve the navy's weather analysis and prediction capabilities and to develop techniques for assessing effects of atmospheric conditions on shipboard, airborne, and land-based communication and weapon systems.

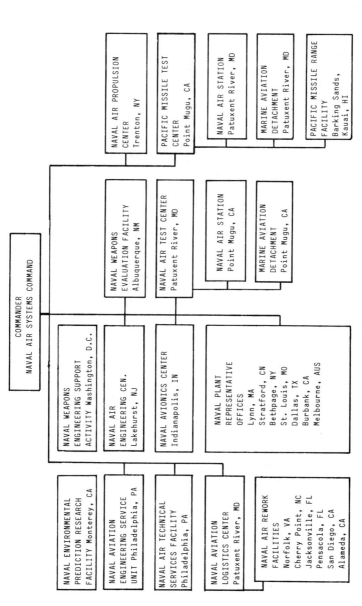

Fig. 12-2. Field Activities of the Naval Air Systems Command

Naval Aviation Engineering Service Unit (NAESU)

This facility, in Philadelphia, provides naval aviation activities with field engineering assistance and instruction in the installation, maintenance, repair, and operation of all types of aviation systems and equipment. NAESU gives the same support to the defense security programs/Foreign Military Sales (FMS).

Naval Air Technical Services Facility (NATSF)

Located in Philadelphia, NATSF manages the NAVAIR technical manual program and develops and implements concepts and policies related to technical manual planning, monitoring, budgeting, procurement, production, quality assurance, reproduction, distribution, and the review and evaluation of delivered products.

Naval Aviation Logistics Center (NAVAVLOGCEN)

Located on the Patuxent River in Maryland, this center is the commander's representative on matters of in-service aviation-weapons maintenance and material support. The center coordinates fleet logistics requirements with depot rework facilities.

Naval Air Rework Facilities (NARF)

The facilities perform depot-level maintenance on systems and equipment. NARFs have the industrial capability to manufacture parts, and they have the technical capability for engineering and design changes. They can perform all levels of aircraft maintenance. The NAVAVLOGCEN is both ISIC and PSO for all six of the navy's NARFs. They are located in Norfolk, Virginia; Cherry Point, North Carolina, which is primarily used for marine corps aircraft; Jacksonville, Florida; Pensacola, Florida; Alameda, California; and North Island at San Diego, California.

Naval Weapons Engineering Support Activity (NAVWESA)

Located in Washington, D.C., this activity may be called NAVAIR's center of expertise for the production of weapon systems. NAVWESA acquires the items of government-furnished equipment necessary for the production of weapon systems. It also reviews the adequacy of technical data used in the manufacture of weapon systems and their related equipment, and it maintains acquisition information systems.

Naval Air Engineering Center (NAEC)

This center, in Lakehurst, New Jersey, conducts programs of research, engineering development, developmental testing and engineering, systems integration, limited production, procurement, and fleet engineering support. The programs are used for the following equipment: aircraft launching and recovery systems, landing aid systems, and support equipment for aircraft and airborne weapon systems.

These Lockheed P-3 Orion aircraft are being overhauled and fitted with the latest modifications at the naval air rework facility in Alameda, California. (Merco International)

Naval Avionics Center (NAC)

The activities of this center, located in Indianapolis, are concerned with research and development, engineering, material acquisition, pilot and limited manufacturing, technical evaluation, depot maintenance, and integrated logistics support of airborne electronics, missile weapon systems, and spaceborne weapon systems.

Naval Plant Representative Offices (NAVPRO)

These offices are established at major contractors' plants. NAVPROs check to see if company procedures, materials, engineering, industrial management, and production all conform with contractual requirements. There are seven NAVPROS:
  —Sikorsky Helicopter Plant in Stratford, Connecticut
  —Grumman Aircraft Plant in Bethpage, Long Island
  —Lockheed Aircraft Plant in Burbank, California
  —McDonnell Douglas Aircraft Plant in St. Louis

A jet engine is put through its paces at the Naval Air Propulsion Center, Trenton, New Jersey. (U.S. Navy)

—Vought Aircraft Plant in Dallas
—General Electric Engine Plant in Lynn, Massachusetts
—One in Melbourne, Australia, for Australian purchases of the F/A-18 Hornet

Naval Air Propulsion Center (NAPC)

This center, in Trenton, New Jersey, provides NAVAIR and the fleet with complete technical and engineering support for air-breathing propulsion systems. It manages and also performs applied research and development of new propulsion systems. It helps develop and evaluate new propulsion systems and tests and evaluates propulsion systems.

Naval Weapons Evaluation Facility (NWEF)

This facility, at Kirtland Air Force Base in Albuquerque, New Mexico, tests, evaluates, and supports nuclear and nonnuclear weapons. The facility liaises with all levels of command within the navy and other government agencies regarding nuclear weapon

Test bombing results are carefully scrutinized at the naval weapons evaluation facility at Kirtland Air Force Base, Albuquerque, New Mexico. (U.S. Navy)

safety. It plans and coordinates the navy nuclear-weapons safety program. It assists in the trials of naval aircraft when requested to do so by the Board of Inspection and Survey.

Naval Air Test Center (NATC)

This center, on the Patuxent River, tests and evaluates aircraft weapon systems, components, and related equipment to determine their suitability for service use. The center is ISIC and PSO for the naval air station and the marine aviation detachment at Patuxent River.

Pacific Missile Test Center (PMTC)

This center, at Point Mugu, California, tests and evaluates naval weapons systems and provides developmental, engineering, logistics, and training support for them. It furnishes the operating forces and other government agencies with major range, technical, and base support. PMTC is the ISIC and PSO for the naval air station and marine aviation detachment at Point Mugu and for the Pacific missile-range facility, Barking Sands, Kauai, Hawaii.

13

The Naval Air Reserve

By Commander Peter B. Mersky, USNR

Readiness is the hallmark of any good military organization, and readiness is what the naval air reserve is all about. The purpose of the reserve is to maintain a civilian force of qualified aviation officers and enlisted personnel who are trained and prepared for mobilization and immediate action. The reserve is expected to keep abreast of changing technology and to maintain its proficiency so that in time of national emergency it can respond to any contingency. In a large-scale conflict, it must be able to augment the existing active-duty force to meet the expanding needs of naval aviation while new personnel are being trained.

Pilots, NFOs, and enlisted technicians are expensive commodities. The investment in both time and money for training aviation personnel and maintaining their operational skills is considerable, and it is lost when personnel are released from active duty. The cost, however, of maintaining a large active-duty force in time of peace is prohibitive. The solution to the problem, and one that

The Naval Air Reserve (U.S. Navy)

The naval air reserve maintains a civilian force of qualified personnel who can be called upon at a moment's notice in wartime or crisis situations. (Peter Mersky)

has proven itself on numerous occasions, is the maintenance of a naval air reserve.

HISTORICAL PERSPECTIVE

Until World War I, the only naval reserve forces were the naval militia of various states. Naval militias were originally sponsored by civilian amateur sailors and naval enthusiasts rather than by federal or state governments. As naval militias proved effective, some states began to purchase clothing and equipment to help defray the expense to the members of the militia organizations.

In 1891, as part of a naval appropriations act, Congress allocated twenty-five thousand dollars for arming and equipping naval militias. Ships were loaned to these organizations, and training assistance was given by the bureaus of the Navy Department. The program was administered by the Office of the Naval Militia. With annual practice cruises, training gradually progressed.

During the Spanish-American War of 1898, the naval militia proved its value and efficiency in time of national emergency. Because of this, the Navy Department recommended the organization of a national naval reserve. In February 1914 the Naval Militia Act became law. All states were required to organize their naval militia units in consonance with the Navy Department's prescribed plans.

In April 1914 plans were initiated for the establishment of an aeronautical force in the naval militia, and the following year the navy's General Order 153 established the practice of lending planes to the states that had aeronautical corps in their naval militia.

The state of New York established an aviation militia at Bay Shore, Long Island, in May 1911. In July of the same year, the Massachusetts militia, including an aeronautical branch at Squantum, was mustered into federal service. The field and facilities were supplied by local authorities. The Aero Club of America assisted in the administration of the program. This group offered the services of many of its members to the militia and sought ways to contribute to the meager funds of the navy for the purchase of aircraft.

The naval appropriations act for fiscal year 1917 provided funds for the establishment of a naval flying corps and the purchase of twelve planes for the naval militia. The naval reserve flying corps was designed to attract civilian aviators, designers, and aircraft workers. It also attracted many college student groups. The first of these was the "Yale unit," organized in 1916 by F. Trubee Davison, who later became assistant secretary of war. (His name is on the annual trophy given to the best naval air reserve squadrons in various categories.) The organization of the First Yale Unit is considered the official beginning of the naval air reserve.

In 1917, after the United States declared war on Germany, the commandant of the First Naval District assumed control of the naval militia station at Squantum, Massachusetts, across the harbor from Boston, for use as an air training station. This was one of several actions taken immediately after the declaration of war to expand the air training program while more permanent bases were being built. Other units brought under the control of the navy included the naval militia station in Bay Shore, Long Island; the Curtiss Flying School in Newport News, where a Harvard unit was in training; the Goodyear Balloon and Dirigible Training Facility in Akron, Ohio; a Princeton unit in East Greenwich, Rhode Island; and the two Yale units at West Palm Beach, Buffalo, and Long Island. During World War I, the four thousand officers and twenty thousand enlisted men of the naval reserve flying corps accounted for three-fourths of the overall aviation strength.

By mid-1922 the large war component of the naval reserve force had been reduced to practically nothing. Early in 1923, Rear Admiral William A. Moffett, chief of the Bureau of Aeronautics, submitted to the CNO a plan to correct deficiencies in the naval air reserve. He recommended that at least one naval air reserve unit be located in each naval district, and on 13 August Squantum, Massachusetts, became the first organized naval air reserve establishment. In 1926 Admiral Moffett instigated a plan to raise the number of naval reserve aviation units to seven. In July 1927 the first group of fifty newly commissioned ensigns of the naval aviation reserve was ordered to one year of training duty with the fleet. The Naval Air Reserve continued to expand and, immediately prior to World War II, consisted of sixteen activities spread throughout the United States.

In October 1940 the secretary of the navy placed all divisions and aviation squadrons of the organized reserve on short notice for recall. By January 1941 all units had been ordered to active duty. During World War II members of the naval air reserve took their place beside their regular navy counterparts and made significant contributions to the final victory.

After World War II, the Naval Air Reserve Training Command, with twenty-one activities, was formally activated as a component of the Naval Air Training Command.

Fourteen squadrons of the organized reserve were activated in July 1950 for duty in the Korean conflict. By the end of this operation over forty reserve squadrons had received mobilization orders. The wing of one attack carrier, made up entirely of recalled reservists, established an impressive record in combat.

As the naval air reserve grew, up-to-date fleet training and equipment was clearly needed to maintain reservists in a state of readiness. In November 1957 reserve squadrons first began firing guided missiles as part of their regular training. In 1959 units of the naval air reserve participated for the first time in a full-scale fleet exercise. In July of that year the navy announced that reserve squadrons would receive Sidewinder air-to-air missiles.

The Berlin Crisis in 1961 saw President John F. Kennedy call up eighteen selected air reserve squadrons. This was the first partial mobilization of selected reservists.

In the spring of 1965 the commander in chief of the U.S. Pacific Fleet asked for the support of the Naval Air Reserve in logistics operations in Southeast Asia. The voluntary response of the selected air reservists, who delivered vital air cargo to Saigon and Da Nang, was overwhelming.

The naval air reserve held a nationwide celebration of its fiftieth anniversary in 1966.

In January 1968 the air reserve was again called upon in response to the Pueblo Crisis. Six VA/VF squadrons were recalled to active duty; however, because of training and equipment deficiencies, these units were put on inactive duty.

As a result, during 1970 the naval air reserve was reorganized into a mirror image of the active fleet squadrons. Two reserve attack-carrier air wings, two reserve helicopter antisubmarine groups, and two fixed-wing reserve patrol wings were created. In addition, the transport squadrons were reorganized into units strategically placed around the country.

On 1 February 1973 the Naval Air Reserve Command in Glenview, Illinois, consolidated with the Naval Surface Reserve Command, then located in Omaha, Nebraska, to form the office of the chief of naval reserve, headquartered in New Orleans, Louisiana.

ORGANIZATION

The naval air reserve and its programs are directed by the deputy chief of naval reserve (air)/commander naval air reserve force (COMNAVAIRESFOR). This command is under operational and administrative control of the chief of naval reserve. Under COM-

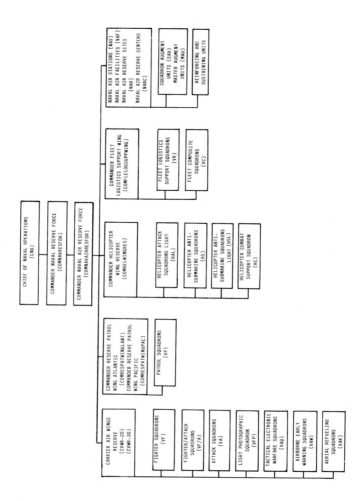

Fig. 13-1. Command Organization of Commander, Naval Air Reserve Force

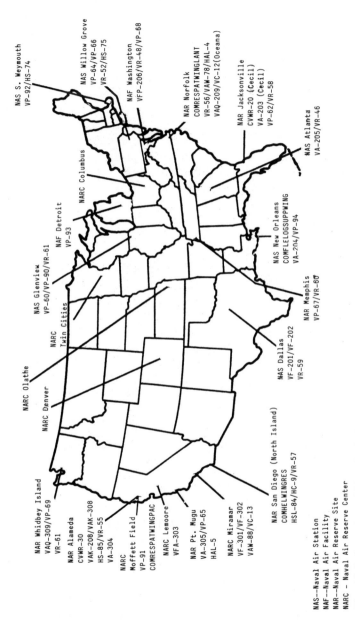

NAS S. Weymouth
VP-92/HS-74

NAS Willow Grove
VP-64/VP-66
VR-52/HS-75

NAF Washington
VFP-206/VR-48/VP-68

NAR Norfolk
COMRESPATWINGLANT
VR-56/VAW-78/HAL-4
VAQ-209/VC-12(Oceana)

NAR Jacksonville
CVWR-20 (Cecil)
VA-203 (Cecil)
VP-62/VR-58

NAS Atlanta
VA-205/VR-46

NARC Columbus

NAF Detroit
VP-93

NAS Glenview
VP-60/VP-90/VR-61

NARC
Twin Cities

NARC Olathe

NARC Denver

NAS New Orleans
COMFLELOGSUPPWING
VA-204/VP-94

NAR Memphis
VP-67/VR-60

NAS Dallas
VF-201/VF-202
VR-59

NAR Whidbey Island
VAQ-309/VP-69
VR-61

NAR Alameda
CVWR-30
VAK-208/VAK-308
HS-85/VR-55
VA-304

NARC
Moffett Field
VP-91
COMRESPATWINGPAC

NARC Lemoore
VFA-303

NAR Pt. Mugu
VA-305/VP-65
HAL-5

NARC Miramar
VF-301/VF-302
VAW-88/VC-13

NAR San Diego (North Island)
COMHELWINGRES
HSL-84/HC-9/VR-57

NAS--Naval Air Station
NAF--Naval Air Facility
NAR--Naval Air Reserve Site
NARC - Naval Air Reserve Center

Fig. 13-2. Location of Naval Air Reserve Activities

NAVAIRESFOR are the reserve wings and squadrons, air stations and facilities, and training sites and centers.

Each aircraft type and the responsibility for it is allocated to either a Pacific or an Atlantic mobilization billet. For example, in the event of mobilization, CVWR-20, one of the reserve attack carrier wings, would report to the Atlantic Fleet, while CVWR-30 would deploy to the western Pacific. There is one reserve helicopter wing, which includes four antisubmarine units, one combat search-and-rescue squadron, and two squadrons for light-helicopter attack duties. Eleven squadrons of the support wing provide fleet logistic support, while two composite squadrons undertake a variety of missions, including adversary and target-towing services to the fleet. In all, the naval air reserve has some fifty-one squadrons operating over four hundred aircraft. (See appendix C for reserve-squadron locations.)

There are six naval air stations dedicated to naval air reserve activities: Dallas, New Orleans, Glenview (near Chicago), Atlanta, Willow Grove (near Philadelphia), and South Weymouth (near Boston). Naval air facilities, naval air reserve units, and naval air reserve centers are usually tenant commands of larger military installations.

The major naval air reserve activities are located within the major population centers of the United States to eliminate many problems associated with extensive travel to weekend drill sites.

There are some twenty-two thousand naval air reservists, which accounts for about 17 percent of the navy's total aviation force. The naval air reserve provides the following percentages of the navy's overall force:

—14 percent of tactical carrier air wings
—33 percent of composite air wings
—35 percent of maritime patrol wings
—100 percent of light-attack helicopter squadrons
—100 percent of combat search-and-rescue capability
—100 percent of U.S.-based heavy logistic airlift

AUGMENT PROGRAMS

Squadron augment units (SAUs) and master augment units (MAUs) maintain a high degree of readiness, and their personnel are available to augment existing fleet units in the event of mobilization. SAU personnel fly with a reserve squadron or with a regular fleet replacement squadron. MAU personnel train with regular active-duty squadrons but have their own aircraft.

OTHER AUGMENT ACTIVITIES

In addition to the selected reservists in SAUs and MAUs, other air reserve personnel are assigned to various reinforcing and sustaining units, which will augment existing regular commands and activities during a mobilization. These units include:

—Aircraft carrier units
—Base and station units
—Air staff units
—Tactical aircraft squadrons
—Antisubmarine warfare operations centers
—Oceanographic units

The naval reserve provides all of the navy's U.S.-based heavy logistic support. (U.S. Navy)

—Air Systems Command units
—Aviation supply units
—Audiovisual units

RESERVE TERMINOLOGY

Like any other large organization, the naval air reserve has its own jargon. Some of the more commonly used terms are discussed below.

Active duty for training (ACDUTRA) means full-time training duty with a regular component of the navy, usually for a period of twelve to fourteen days.

An augmentation unit is normally a squadron of replacement pilots and personnel from which the regular squadrons draw when there is a vacancy. Augmentation units fly and help maintain aircraft.

The ready reserve comprises all reservists liable for active duty in time of war or national emergency. Reservists in this category are in an active status; only ready reservists may receive pay for taking part in inactive-duty training.

The retired reserve are reservists on the retired list who can be ordered to active duty only in time of war or national emergency. Retired reservists include members retired with or without pay.

The selected reserve are ready reservists in a drill-pay status. Selected reservists are issued mobilization orders in time of peace and are to report to specified activities in time of war or national emergency.

A selected air reservist is a member, officer or enlisted, of a naval air reserve squadron. The term specifically pertains to reservists in a drill status.

TAR stands for the training and administration of the reserve. TARs are on active duty with special designators and are involved with the naval reserve training program.

AFFILIATION

Individual pilots and NFOs who wish to affiliate with the naval air reserve can do so through the reserve activity nearest their active-duty station, their residence, or their intended residence. Active-duty personnel can apply ahead of time for immediate assignment to a naval air reserve unit upon release from active duty. This eliminates administrative delays in training and pay.

The youngest, best-qualified officer has priority for assignment to pay billets in the naval air reserve program. These pay billets are limited in number by regulation.

TRAINING

Normally all naval reservists attached to drilling units are expected to attend a minimum specified number of drills per year—usually forty-eight—and one twelve- to fourteen-day period of annual active duty for training. A drill weekend usually consists of four four-hour drills. There are two categories of drills, pay and nonpay. Naturally, the pay billets are the most sought after, for several thousand dollars a year can be earned by a lieutenant or lieutenant commander in a pay billet.

Readiness is the key to an effective naval air reserve. Here a Mc-Donnel Douglas F-4 Phantom II of Fighter Squadron 202, based at the naval air station in Dallas, catches a wire aboard the USS *Eisenhower*. (U.S. Navy)

AIR RESERVE UNITS

The naval air reserve utilizes all the types of aircraft in the fleet. Reserve squadrons and wings are organized like their fleet counterparts to make integration as smooth as possible. There are attack, fighter, early warning, electronic, antisubmarine patrol, fleet tactical support, fleet composite, photoreconnaissance, and helicopter squadrons. Others include mobile support, base support, and various training units, many of which augment their active-duty fleet sponsors in the event of mobilization. Major air stations, for example, have reserve augmentation units, as do aircraft carriers. The members of these units usually train with the fleet sponsor.

Some reserve naval aviators serve as flight instructors with the Naval Air Training Command during their active-duty training. Some reservists drill and fly with training-command activities in selected training areas and squadrons.

Many reservists are willing and able to devote more than the required amount of time to their reserve responsibilities. Additional drill and active-duty opportunities are frequently offered to keep skills honed.

TANGIBLE BENEFITS

Reservists in a drill-pay status receive one day's base pay plus flight pay (if eligible) for each drill performed. Normally, two drills are held on Saturday and two on Sunday, one weekend a month. Thus, the naval air reservist is paid four days' pay for each weekend of drills. Added to this is the pay for any additional drills. Only basic pay and flight pay are earned during the regular weekend and additional drills. Full pay and allowances are earned during the annual two-week active-duty cruise and for any other periods of temporary active duty.

Most squadrons in the naval air reserve have at least one additional period of active-duty training when time allows and needs demand. For the fighters and attack units, these extra stints usually involve short detachments to ranges or competitions to further hone skills. Patrol squadrons may find themselves deployed to various bases.

Reserve pilots maintain proficiency during monthly drills and annual periods of active duty for training. Many reservists devote additional time to keep skills current. (U.S. Navy)

PROMOTION

Naval air reservists are eligible for promotion all the way up to rear admiral. Selection procedures are similar to those used for active-duty personnel.

Reserve selection boards are composed of active-duty and in-active-duty officers. The boards select those reserve officers best qualified for promotion. As with active-duty personnel, reserve officers are selected primarily on the basis of fitness reports.

RETIREMENT

Members of the naval air reserve become eligible for retirement pay after completing twenty years of "qualifying service." In order to be credited with a year of qualifying service, the naval reservist must accumulate a minimum of fifty retirement points in an anniversary year.

Retirement points are earned as follows:
—One point for each day of active duty or active-duty training (including travel time)
—One point for each authorized drill attended in either a pay or nonpay status
—Points for the completion of approved correspondence courses, the number depending upon the course completed
—Fifteen points credited for each year of active-status membership in a naval reserve component

Retirement pay starts at age sixty, provided the retiree, during the last eight years of qualifying service, served as a member of a reserve component. These eight years need not be continuous, nor is it necessary that participation for the last eight years be in a drilling unit. A reservist is not eligible for retired pay if he is eligible for or receiving any other retired pay for military service.

The rate of retired pay is based on a formula that takes into account the total number of retirement points accrued. A simple and reasonably accurate method of estimating the rate is to credit 2.5 percentage points for each year of active duty and 0.5 percentage points for each year of inactive duty. The combined percentage, multiplied by the basic pay for the grade in which one retired, gives the approximate monthly retired pay. Maximum retired pay cannot exceed 75 percent of such basic pay. Social security and civil service retirement pay may be received concurrently with navy retirement pay.

After retirement with pay, the reservist is entitled to use armed forces facilities subject to availability of space. Included among these are exchanges, commissaries, package stores, and recreational facilities. Medical care is also available for the retiree and eligible dependents.

OTHER BENEFITS

The aviator benefits in many other ways by participating in the naval air reserve program. Besides being able to remain involved in naval aviation, he is able to use most of his base's recreational facilities and shop in the exchange (though not the commissary). On his two-week annual cruise, the reservist has full military ex-

The naval air reserve flies most of the aircraft found in the active navy. Seen here are Grumman E-2 Hawkeyes of Carrier Airborne-Early-Warning Squadron 88. (U.S. Navy)

change and commissary privileges as well as medical care if he sustains an injury or becomes sick during that period. If the reservist dies while in a duty status, survivor benefits are given to dependents. There is also the perquisite of low-cost group-term life insurance, which can be automatically deducted from the monthly paycheck.

Perhaps the greatest benefit derived from participation in the naval air reserve is the somewhat intangible reward of pride and satisfaction that comes with serving one's country.

THE TAR PROGRAM

Operating and maintaining a naval air reserve base, squadron, or other activity is a full-time affair, and one that does not begin and end with weekend drills or two-week cruises. The drilling reservist cannot be with his unit throughout the week, so a special kind of reservist called a TAR deals with day-to-day operations and provides continuity.

Reservists of Helicopter Antisubmarine Squadron 128 perform active duty for training aboard the USS *Constellation*. (U.S. Navy)

TARs are the nucleus of the rest of the unit. They are indispensable links between the weekend warriors and the unit. TAR duties typically include training, organizing, recruiting, instructing, and administering the naval reserve component.

Reserve officers on active or inactive duty in the unrestricted line and supply corps, in the grade of lieutenant (j.g.), lieutenant, or temporary lieutenant commander, are eligible for the TAR program. Officers must have completed their initial active-duty obligation to be eligible. Applications are considered quarterly by a TAR selection board, which evaluates the applicant's previous experience, performance, and qualifications as well as the needs of the service.

TAR officers compete only among themselves for promotion; they are a competitive category of officers with similar background and experience.

A UNIQUE EXPERIENCE

In some respects, the responsibilities of a naval reservist are greater than those of his active-duty counterpart. A reservist holds two jobs, and to keep them he must satisfy two bosses, one a civilian and one a military man. The naval air reserve aviator must keep himself current in all aspects of the operation of his particular aircraft and its mission. It is a demanding regimen that an active reservist sets for himself.

Naval air reservists are highly motivated. Most naval aviators of World War II were reservists. Significant numbers of reserve aviators were on hand again in Korea and Vietnam. More recently, reservists flying C-9 transport aircraft were called upon on short notice to evacuate casualties from Beirut. Reserve fighter squadrons deploy overseas to work with carrier forces while patrol squadrons fill in for active-duty squadrons in the Atlantic and Pacific. Still other reservists flying E-2 and P-3 aircraft help track down drug smugglers trying to slip their illegal cargoes into the United States.

Whatever a reservist's civilian occupation, when he is on duty he is a dedicated navy professional doing a professional job.

14

Marine Corps Aviation

By Major John M. Elliot, USMC (Ret.)

HISTORY

Since the establishment of marine corps aviation, all marine corps aviators have been designated naval aviators. The first marine corps aviator was First Lieutenant Alfred A. Cunningham. Cunningham entered the marine corps in 1909 after serving in the army during the Spanish-American War. His early interest in aviation and his constant requests for aviation duty resulted in orders to report to the aviation camp at the Naval Academy on 22 May 1912. This date has come to be recognized as the birthday of marine corps aviation. Cunningham, upon soloing, was designated naval aviator no. 5 and marine corps aviator no. 1.

The third aircraft acquired by the navy, the B-1, was assigned to Cunningham. It had been wrecked several times and was in very poor condition. In a letter to Captain Chambers, Cunningham complained of vibration in the aircraft and of its inability to climb

Marine Corps Aviation (U.S. Navy)

over a few hundred feet with a passenger. Eventually Cunningham rebuilt the plane.

Progress was slow in naval aviation, primarily because of a limited amount of equipment and personnel. During the early years, however, several operations were carried out with the fleet which included reconnaissance patrols and early shipboard catapult testing. During operations at Tampico and Vera Cruz, marines flew in support of the landing parties.

Flying in those early days was primitive, and men felt their way gingerly into the aviation environment. One of the more colorful contributions of that trial and error period was made by Captain Francis T. Evans at Pensacola on 13 February 1917. Previously it was thought impossible to loop a seaplane because of its heavy pontoon. After several attempts Evans found that he was able to perform the maneuver. Then, to be sure he had witnesses, he performed it again over the hangar area. In his attempts to loop, the N-9 fell off into a spin several times. When he moved the controls to bring the aircraft to heel, he became the first American pilot to recover from a spin in a seaplane. These feats were eventually recognized in 1936, when he was awarded the Distinguished Flying Cross.

When the United States entered World War I, a rapid buildup in marine corps aviation took place, as it did in all branches of the services. The Marine Aeronautic Company was organized as a component of the Advance Base Force at the marine barracks, Philadelphia Navy Yard. Within six months the company was divided into the First Aviation Squadron, which flew land planes, and the First Marine Aeronautic Company, which flew seaplanes. Under the command of Captain Evans, the latter group shipped overseas on 9 January 1918. It was the first fully equipped and trained American aviation unit to go overseas. Equipped with N-9s, R-6s, and HS-2Ls, the squadron conducted antisubmarine patrols from Ponta Delgada in the Azores for the duration of the war.

The land-plane squadron, commanded by Captain William McIlvain, moved to Mineola, Long Island. Due to the extreme cold of the 1917 winter, it moved south to Lake Charles, Louisiana, where it was incorporated into the army training program. A new unit known as the aeronautical detachment was formed at Philadelphia in December. Early in 1918 the aeronautical detachment, under command of Captain Roy Geiger, moved from Philadelphia to Miami. Located first at the main navy base of Coconut Grove, he soon moved to a small strip operated by Glenn Curtiss on the edge of the Everglades. He acquired not only the base but also the aircraft, and a number of the personnel were induced to enter the service. On 1 April McIlvain's squadron arrived from Lake Charles.

Cunningham was now acting as the director of marine aviation as well as the CO of the First Aviation Force, as this Miami-based unit was now known. A vigorous recruiting campaign was conducted at the officers' school at Quantico, and numerous enlisted men also became part of marine corps aviation. This, however, was not enough to flesh out the four squadrons. From the navy training fields several naval officers who wanted to go to France were recruited. Most of these were young reservists already qualified as seaplane pilots. They resigned from the navy, signed up with the marines, and reported to Miami for training in land planes.

On 16 June 1918 the Miami operation was organized into a headquarters detachment and four squadrons, designated A, B, C, and D and known collectively as the First Marine Aviation Force. On 10 July orders were received to embark for France. The force, less Squadron D, departed from New York three days later. The Miami field continued to expand and train Squadron D and other personnel.

In France the First Marine Aviation Force became the day wing of the Northern Bombing Group. This is the first time the wing and group designations were used in the marine corps, even though in reverse order of what we use today. Lack of equipment plagued the force and many marines flew temporarily with British squadrons while awaiting aircraft. The British had an excess of De Havilland 9 (DH-9) airframes, while the Americans had an excess of Liberty engines. McIlvain, who was able to "horse trade" three Liberty engines for a complete DH-9 with a Liberty engine, soon began sending these aircraft to the squadrons in France from the depot in England. Some thirty-six De Havilland aircraft (twenty DH-9As and sixteen DH-4s) were acquired in this fashion. On 5 October the force was brought up to full strength by the arrival of Squadron D from Miami.

First Lieutenant Mulcahy and Corporal McCullough were the first to shoot down a German aircraft (Fokker D-VII) over Coresmarch, Belgium, while flying with Royal Air Force Squadron 218 in a DH-9. The marines' first aerial resupply mission took place when four marine aviators delivered 2,600 pounds of food and stores to a French infantry regiment isolated and surrounded by the enemy.

By mid October 1918 the marines had enough aircraft and had begun to fly missions on their own. On the fourteenth a flight of eight DHs dropped 2,218 pounds of bombs on the railroad yards at Thielt, Belgium. Lieutenant Ralph Talbot and Corporal Robert G. Robinson each received the Medal of Honor for this mission.

While the First Marine Aviation Force was small, its members compiled an enviable record in the short period between 9 August and 11 November 1918. They participated in forty-three missions with the British and conducted fourteen of their own, dropping a total of 33,932 pounds of bombs. Four pilots were killed, one pilot and two rear-seat gunners were wounded. Four confirmed enemy fighter aircraft were shot down and an additional eight were claimed. Not a bad record for a two-seat bomber against German fighters. Lieutenant Talbot was killed in an accident while four officers and twenty-one enlisted men died from the influenza epidemic that wreaked havoc among all the troops in France.

With the war over, marine corps aviation returned to Miami. As with the other services, there was a rapid demobilization and reorganization of the remaining assets. New small squadrons were formed and deployed to Haiti and Santo Domingo to operate with the marine ground forces employed in establishing order in those countries. A squadron was sent to Quantico, Virginia, and another to Parris Island, South Carolina. This squadron later joined the other at Quantico. Flight L was stationed at Sumay, Guam and was the first marine aviation unit in the Pacific. It was this base that was taken over by Pan American when that company established its route across the Pacific.

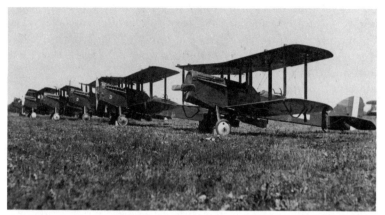

USMC DeHaviland DH-4 aircraft at Yorktown, Virginia, in 1921.
(U.S. Navy)

Once again Cunningham led the way as he continued to develop his concept of the proper employment of the airplane in the marine corps mission. Support of ground troops is still the primary purpose of marine corps aviation and the basis of the tried and proven air-ground team.

Even in those early days, Cunningham visualized the value of radio communication between the aircraft and ground units. He understood the concept of isolating a beachhead by bombing railroads, roads, and reinforcements, and of suppressing beach defenses by bombs and machine-gun fire. All of these ideas were soon to be tried in actual combat operations.

At about this time, Cunningham was replaced by Lieutenant Colonel Thomas C. Turner as the head of marine corps aviation. A vigorous proponent of aviation and a stern disciplinarian, he led marine aviation successfully through its growth period.

To demonstrate the feasibility of long-distance deployment of aircraft, several extended flights were conducted in the early 1920s. Two De Havilland DH-4s, flying from Washington, D.C., to Santo Domingo in 1921, established a record for the longest unguarded (that is, not monitored by surface units) flight over land and water up to that time by U.S. naval aviation. In 1923 four Martin MT bombers were flown from San Diego to Quantico in eleven days for the first mass aerial delivery of aircraft from coast to coast. Again in 1923, a record was established when two DH-4s flew from Port-au-Prince, Haiti, and Santo Domingo City to Washington, D.C., St. Louis, San Francisco and back. They flew 10,953 miles in 127 flying hours over a period of two and a half months. While this may seem a poor showing by those who routinely jet across the country in a commercial airliner, it represented an outstanding achievement for both men and machines in 1923, when there were few airports and equipment was none too reliable.

Aircraft performance during the 1920s and into the 1930s was improved by participation in the great air races of that era. Many racing aircraft were in fact modified service aircraft, and their improved performance during races resulted in follow-on procurements. The marine corps participated in these events with everything from the MT bomber to a biplane fighter that had been converted into a monoplane racer. Cross-country navigation and instrument flying were other activities that helped improve aircraft performance.

Between World War I and World War II the marine corps, including its aviation component, was the only U.S. military service actually to see combat. Early in the decade squadrons in Haiti and Santo Domingo operated against rebel forces there.

The outbreak of civil war in Nicaragua in 1927 led to U.S. intervention and the introduction of marines in an attempt to stabilize the government. This small-scale, drawn-out guerrilla war provided the opportunity to practice some of the theories developed at Quantico and was the beginning of limited close-air support. Dive-bombing, air-ground communications, and aerial delivery of troops and supplies were all utilized in the support of the marine patrols on the ground. The lessons learned in Nicaragua led to doctrines used in the Pacific in World War II and also produced some of the most successful leaders both in the air and on the ground.

Back in the United States the Depression caused a considerable reduction in the strength and consolidation of marine corps aviation throughout the 1930s. Despite this, two squadrons were formed as active components of the new navy carrier force. Personnel were rotated through these squadrons, so that by the time they were deactivated, three years later, two-thirds of the marine aviators had served aboard ship. Marines were not to have regular carrier-based squadrons again till 1945.

One means of bringing marine aviation to public attention was by giving air shows. VF-9M was charged with the task of performing at all the major air shows as well as filling a constant flow of requests for additional appearances. Flying their F4B-4 fighters, VF-9M pilots became widely known for their spectacular aerial displays, in which the entire squadron of eighteen aircraft participated. In their regular military training functions, the members of the squadron were rated at the top of naval fighter squadrons.

A potential source of personnel for a rapid buildup in the event of an emergency was the reserve force, similar to that instituted by the navy. During World War I most marine aviators held reserve commissions. The reserve organization was inactive from 1918 to 1928, at which time five reserve aviators were recalled to active duty. After a brief refresher course, they reported to naval reserve aviation bases to organize marine reserve aviation units. The units were established at Sand Point, Washington; Rockaway Beach, Long Island, Squantum, Massachusetts; and Great Lakes, Illinois. Applicants were selected by COs of the reserve units for pilot training, and those qualified went on active duty for flight training. Those who completed were commissioned second lieutenants, USMCR, and later returned for a one-year active tour and advanced training. Upon their release from active duty they were required to report twice a month at their own expense to the nearest naval reserve base to retain flying proficiency. This quick cycle was one way to obtain a large number of trained personnel in a relatively short time. None of the marine squadrons had their own aircraft; they had to share with navy reserve squadrons who operated on the same bases. Enlisted personnel were trained at navy schools in various aviation specialties. Thirteen reserve squadrons were mobilized and integrated into the regular marine corps at the beginning of World War II.

The expansion of marine aviation just before World War II made the existing system of numbering squadrons impractical. The present three-digit squadron designation was approved on 28 March 1941. This system, as originally conceived, indicated the wing and group to which the squadron was assigned as well as its numerical sequence within the group. However, the great reduction in force at the end of the war and the constant transfers of squadrons have destroyed the original significance of the squadron number.

The attack on Pearl Harbor on 7 December 1941 virtually wiped out marine aircraft assets in the Pacific, although a few planes remained in San Diego, at Wake Island, and on the USS *Lexington*. East Coast squadrons were scattered on maneuvers but within seven days returned to Quantico, packed up, and were on their way to the West Coast. The valiant defense of Wake Island was supported by only twelve aircraft, seven of which were destroyed on the ground during the first attack. Flying new and strange

A Grumman F-4F Wildcat fighter at Henderson Field on Guadalcanal during the Pacific war. Marine corps aviation, fighting against great odds, held the line here and provided a foothold from which to launch an offensive and push the enemy back up the Solomon Islands chain. (U.S. Marine Corps)

aircraft, with a lack of trained mechanics and technical manuals, however, they managed to destroy eight Japanese aircraft before their remaining planes were put out of commission. All squadron members then joined the defense-battalion marines and continued to fight until Wake was captured.

Marine corps aviation came into its own during the U.S. capture and defense of Guadalcanal. Operating on a shoestring, with the outcome often in doubt, the marines were able to beat back the best the Japanese could muster. Thus the long road began up the Solomon Island chain and through other islands toward the Japanese homeland. In the beginning, marine aviators were primarily engaged in an air war, but as the fighting progressed, more effort was devoted to the development of air-ground tactics, though not necessarily in support of the ground marines. Money and adequate equipment had at last become available to put long-held ideas into practice.

Toward the end of 1944, a program was initiated to put marine squadrons on escort carriers (CVEs). A desperate need for additional carrier assets also resulted in ten fighter squadrons serving aboard five carriers of the fast-carrier force. Carrier-based marines flew missions against French Indochina, the Philippines, Formosa, China, and Japan itself.

By the end of the war, marine aviation had grown to five air wings, twenty-nine groups, 132 tactical squadrons, and 116,628 personnel. But with the end of hostilities and the reduction in force, marine aviation phased out the dive-bomber, torpedo bomber, medium bomber, and observation squadrons, retaining only fighter and transport capability. Some units participated in the occupation of Japan and in the attempt to prevent a Communist takeover in China. A few squadrons continued to be the air complement of escort carriers. Conventional squadrons became all-weather qualified; helicopters entered marine aviation in 1947, jet aircraft in 1948, and attack aircraft in 1950. The new equipment and the new techniques were soon to be tested in combat.

The marine air reserve program was reinstated early in 1946 at twenty-one naval air stations. The program grew steadily, so that when the Korean War began it was possible to mobilize rapidly and call personnel to active duty from twenty squadrons to augment the regular forces.

When the North Korean army crossed the thirty-eighth parallel and invaded South Korea on 25 June 1950, Marine Fighter Squadron 214 (VMF-214) was on board the USS *Rendova* (CVE 114) for maneuvers in the Hawaiian area. A request for a marine regimental combat team on 2 July resulted in the designation of Marine Air Group 33 (MAG-33) as the aviation component and the return of VMF-214 to the marine corps air station in El Toro. After a feverish period of activating reserve squadrons and bringing the regular units up to strength, the group departed for the Far East. The first marine air strike was conducted by VMF-214 against the North Korean army on 6 August 1950. Marine squadrons continued to serve on carriers as well as ashore for the duration.

Vought F-4U Corsairs of Marine Aircraft Wing 1 prepare to launch from a carrier during the Korean War. (U.S. Marine Corps)

The concept of vertical envelopment using helicopters, which had been conceived at Quantico, was tried and proven in Korea. Casualty rates were reduced by rapidly evacuating the wounded to rear medical facilities, combat units were lifted into place, and pilots were rescued.

Since the end of the Korean conflict, marine aviation has remained on station in the Far East.

Despite the success of helicopters in Korea, it was still not possible to integrate these aircraft fully into amphibious landing operations. To do so required vessels similar to aircraft carriers but designed to support helicopter and other amphibious operations. As an interim measure, such ships were developed from carriers being phased out of service and were designated landing platform helicopters (LPHs). The first ship designed and built specifically for this task was the USS *Iwo Jima* (LPH 2), commissioned in August 1961.

The need for marine aviation to operate in forward areas where few or no facilities existed had been recognized early in World War II. But it was not until 1958 that a short airfield for tactical support (SATS) was constructed. This installation consisted of an aluminum plank mat and catapult and arresting equipment that could be transported easily and constructed in a minimum amount of time. Three experimental sites were built, and tests continued for five years. In March 1965 the concept was put into combat use with the installation of an SATS at Chu Lai, Vietnam. Eventually, a permanent concrete airfield was constructed further inland.

Naval and marine aircraft and weapons continued to develop. Tactical squadrons all made the transition to jet aircraft. Improved helicopters had significantly greater lifting capability. Marine aviation entered the space age when Lieutenant Colonel John H. Glenn orbited the earth three times in the first U.S.-manned orbital flight.

For marine aviators, involvement in the Vietnam conflict began in April 1962 with the introduction of Marine Medium Helicopter Squadron 362 (HMM-362) as a component of the advisory program. Elements of the First Marine Aircraft Wing began arriving at Chu Lai in April 1965.

Every type of aircraft in the marine corps' operational inventory

Marine corps aviation exists to support marines on the ground. Here marines disembark from a Sikorsky CH-53 Sea Stallion helicopter during an exercise in northern Norway. (U.S. Navy)

was utilized in Vietnam. Tactical squadrons supported troops on the ground and engaged in air-to-air combat. Transport units were kept busy ferrying troops and supplies to and from Japan, providing in-flight refueling for tactical aircraft, and supplying bases such as the besieged Khe Sanh. It was here that the armed helicopter came into its own.

MARINE CORPS AVIATION TODAY

Today's marine corps is still an air-ground team. The role of aviation is to support fleet marine force (FMF) operations by close and tactical air support and defense. A secondary role is to provide replacement or augmentation squadrons for duty aboard aircraft carriers. The noteworthy characteristic of marine corps aviation is that it is an inseparable part of marine corps infantry. Its basic purpose is clear—to support marines on the ground.

ORGANIZATION

The deputy chief of staff for aviation (DC/S) plans and supervises all matters relating to the organization, personnel, operational readiness, and logistic requirements of marine corps aviation. He is also an assistant CNO for marine corps aviation under the DCNO (air warfare). The DC/S for aviation, wearing one hat or another, controls all marine corps aviation, including the organized reserve.

Marine corps aviation is organized into three active-duty aircraft wings and one reserve aircraft wing. The wing is a flexible task organization to which groups, squadrons, and missile units are assigned.

One active-duty marine aircraft wing is located on the East Coast of the United States and another on the West Coast. A third is based in East Asia. For the most part, aircraft squadrons that make up the overseas wing are assigned on a six-month rotational basis from the continental United States and Hawaii.

The Fourth Marine Aircraft Wing is a reserve organization headquartered in New Orleans with aircraft and personnel assets

A ground crewman checks the landing gear of an AV-8 Harrier aircraft prior to takeoff. (U.S. Navy)

located at naval and marine corps air stations throughout the United States.

No two marine aircraft wings are exactly alike, but a hypothetical wing might include a marine wing headquarters squadron (MWHS), a marine air-control group (MACG) with associated antiaircraft missile units, a marine wing support group (MWSG), and three or more marine aircraft groups (MAGs).

The MWHS provides command, administrative, and supply support for an MWHS and certain elements of the MACG, while the MACG coordinates the air command and control system of the wing. The MWSG provides command, control, supply, and logistic support for the squadrons of the wing. It also furnishes such support as motor transport and refueling for ground equipment and aircraft. The MAGs conduct air operations from advanced bases, expeditionary airfields, and aircraft carriers in support of FMFs.

Marine Sea Stallions from the amphibious assault ship *Inchon* (LPH 12) provided support for the Twenty-fourth Marine Amphibious Unit in Lebanon. Beirut is seen in the background. (U.S. Navy)

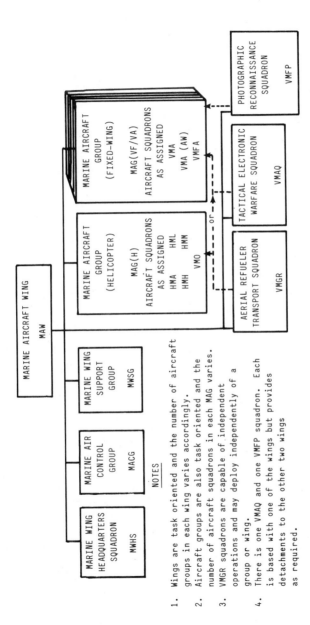

NOTES

1. Wings are task oriented and the number of aircraft groups in each wing varies accordingly.

2. Aircraft groups are also task oriented and the number of aircraft squadrons in each MAG varies.

3. VMGR squadrons are capable of independent operations and may deploy independently of a group or wing.

4. There is one VMAQ and one VMFP squadron. Each is based with one of the wings but provides detachments to the other two wings as required.

Fig. 14-1. Marine Aircraft Wing Organization

MARINE SQUADRONS AND THEIR MISSIONS

Each MAG consists of a tactical command and an administrative support nucleus to which three or more squadrons are attached for operations or training. The squadron is the basic tactical and administrative unit. Its relation to the air group is similar to that of a battalion to a division.

Squadron Type	Letter Designation	Squadron Mission
Attack	VMA	Attacks and destroys surface targets; escorts helicopters and conducts other air operations as directed
All-weather attack	VMA(AW)	Attacks and destroys surface targets under all-weather conditions; escorts helicopters and conducts other air operations as directed
Attack training	VMAT	Conducts combat-capable jet attack training for pilots with emphasis on weapons delivery in destruction of ground targets; provides training for aviation technical personnel
All-weather attack training	VMAT(AW)	Conducts jet attack training for naval aviators with emphasis on the attack and destruction of surface targets under all-weather conditions; provides training for aviation technical personnel
Fighter/attack	VMFA	Intercepts and destroys enemy aircraft under all-weather conditions; attacks and destroys surface targets and conducts other air operations as directed
Fighter/attack training	VMFAT	Conducts combat-capable jet-intercept and fighter/attack training for naval aviators with emphasis upon training for all-weather operations; provides training for aviation technical personnel
Tactical electronic warfare	VMAQ	Conducts airborne electronic warfare in support of FMC operations
Tactical reconnaissance	VMFP	Conducts aerial multisensor imagery reconnaissance in support of FMF operations

Squadron Type	Letter Designation	Squadron Mission
Refueler transport	VMGR	Provides aerial refueling services and assault air transport for personnel, equipment, and supplies
Observation	VMO	Conducts reconnaissance, observation, and air-support operations in support of the FMF
Helicopter attack	HMA	Provides close-in fire support during aerial and ground escort operations during ship-to-shore movement and within an objective area
Light helicopter	HML	Provides utility combat helicopter support to the landing force in the ship-to-shore movement and in subsequent operations ashore
Medium helicopter	HMM	Provides helicopter transport of personnel, supplies, and equipment for the landing force during ship-to-shore movement and within an objective area
Heavy helicopter	HMH	Provides helicopter transport of heavy supplies, equipment, and personnel for the landing force during ship-to-shore movement and within an objective area
Marine helicopter squadron one	HMX	Provides helicopter transportation for the president and vice president, members of the Cabinet, and foreign dignitaries; provides helicopter emergency evacuation as directed by the White House Military Office; provides helicopter support for headquarters marine corps as directed

MARINE CORPS AIR STATIONS

In the continental United States as well as overseas, the marine corps maintains its own air stations and base commands on the stations to support aviation units serving ashore. On each coast there is a marine air base command that commands all marine corps aviation shore establishments within its area. The shore establishments support aviation units of the FMF. The East Coast

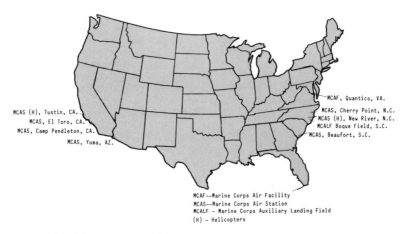

MCAF, Quantico, VA.

MCAS (H), Tustin, CA.
MCAS, El Toro, CA.
MCAS, Camp Pendleton, CA.
MCAS, Yuma, AZ.

MCAS, Cherry Point, N.C.
MCAS (H), New River, N.C.
MCALF Boque Field, S.C.
MCAS, Beaufort, S.C.

MCAF—Marine Corps Air Facility
MCAS—Marine Corps Air Station
MCALF – Marine Corps Auxiliary Landing Field
(H) – Helicopters

Fig. 14-2. Marine Corps Airfields

command is Commander Marine Corps Air Bases, Eastern Area, located at the air station in Cherry Point, North Carolina; that on the West Coast is Commander Marine Corps Air Bases, Western Area, located at the air station in El Toro, California.

MARINE CORPS AIRCRAFT

Marine corps aircraft cover a wide range of types and vintages. Tried and true tactical jets like the A-4 Skyhawk and the F-4 Phantom II continue to serve alongside the revolutionary AV-8 VSTOL (vertical short takeoff/landing) Harriers and the F/A-18 Hornet, whose capabilities are still being developed. Similarly, the UH-1N Iroquois/Huey, the AH-1 Sea Cobra, the CH-46 Sea Knight, and the CH-53 Sea Stallion continue to provide essential services while improved models like the CH-53E Super Stallion and the AH-1T+ Super Cobra are being introduced.

Whatever the function, marine corps aircraft have as their ultimate purpose the support of marines on the ground.

An AV-8B Harrier II takes off with over nine thousand pounds of bombs. The AV-8B is the marine corp's newest V/STOL attack aircraft. (McDonnell Douglas)

A list of marine corps aircraft follows:

Attack aircraft

McDonnell Douglas Skyhawk (A4M, OA-4M, TA-4F, TA-4J)

Grumman Intruder (A-6E, KA-6D)

VSTOL attack aircraft

British Aerospace/McDonnell Douglas Harrier (AV-8A, AV-8B, AV-8C, TAV-8A)

Fighter attack aircraft

McDonnell Douglas Phantom II (F-4S)

McDonnell Douglas Hornet (F/A-18)

Electronic warfare aircraft

Grumman Prowler (EA-6B)

Photo reconnaissance aircraft

McDonnell Douglas Phantom II (RF-4B)

Observation aircraft

Rockwell International Bronco (OV-10A)

Grumman A-6E Intruders from three different marine attack squadrons join up for some close formation flying. (U.S. Navy)

Transport aircraft
  McDonnell Douglas Skytrain (C9-F)
  Lockheed Hercules (KC-130F, KC-130R)
Utility aircraft
  Grumman Academe/Gulfstream (TC-4C)
  Rockwell International Sabreliner (CT-39G)
  Beech Super King Air (UC-12B)
Helicopters
  Bell Sea Cobra (AH-1T)
  Bell Iroquois/Huey (UH-1N, VH-1N, VH-3D)
  Boeing Vertol Sea Knight (HH-46A, CH-46E)
  Sikorsky Sea Stallion (CH-53A, CH-53D)
  Sikorsky Super Stallion (CH-53E)

15

Coast Guard Aviation

By Commander Jess C. Barrow, USCGR (Ret.)

THE BEGINNING

Coast guard aviation officially began on 30 March 1916, when
Second Lieutenant Charles E. Sugden and Third Lieutenant Elmer
F. Stone were selected by the Treasury Department to report to
the naval air station at Pensacola, Florida, for flight training. On
29 August of that year, Congress passed the Navy Deficiency Act,
which authorized the Treasury Department to establish ten coast
guard air stations along the Atlantic and Pacific coasts, the Great
Lakes, and the Gulf of Mexico. The legislation also authorized a
coast guard aviation school and an aviation corps of ten flight
officers, five engineering officers, and forty mechanics. Congress,
however, failed to appropriate the funds to carry out its legislation,
and the program was shelved until the termination of World
War I.

In the meantime, the coast guard sent four additional officers
and twelve enlisted men to Pensacola. All received their wings in

The first aircraft built for the coast guard was this Loening OL-5 amphibian, delivered in October 1926. (U.S. Coast Guard)

1917. During the war, coast guard pilots were attached to the navy and served in most aviation billets. One coast guard officer served as CO of the naval air station at Ille Tudy, France, and was honored by the French government.

After the war, coast guard aviators continued to fly navy airplanes. The Atlantic Ocean was first crossed by a flying boat in 1919. One of the pilots on this memorable transatlantic flight was coast guard Lieutenant Elmer F. Stone.

In March 1920 the first coast guard air station was established at Moorehead City, North Carolina, with six airplanes borrowed from the navy. After fifteen months of operations, the station was closed because of a lack of funds. However, the Moorehead City experiment convinced the Treasury Department that its air arm was a vital necessity.

In 1925 a new air station was established at Gloucester, Massachusetts. It was here that Lieutenant Commander Carl C. von Paulson and Aviation Pilot Leonard M. Melca flew many thousands of miles on patrol, locating rum-running boats, directing patrol boats to vessels in distress, carrying out experiments in two-way radio communications, and performing many other duties. The result convinced Congress to give the coast guard $152,000 to purchase five airplanes of its own. Within a short time the coast guard had contracts for three Loening OL-5 Amphibians and two Chance Vought UO-4 floatplanes.

During the 1930s coast guard aviation grew steadily and rapidly. It became an integral part of the entire coast guard and was designed to carry on the traditional and time-honored duty of the service. Air stations, their number increased to nine, were strategically located to enable the greatest service to the public and the military.

By 1939 the number of cases of coast guard assistance and of hours flown by coast guard pilots had increased tenfold. The magnitude of the flight activity can be seen from the following statistics, amassed in 1939: area of search on regular patrols, 9,300,000 square miles; time in flight, 13,230 hours; 353 assist flights; 223 people assisted; 29,230 vessels identified; 1,466 people warned of impending danger; 259 vessels warned; 133 emergency cases transported; 192 cases of assistance to other government departments; 76 disabled vessels located; and 4,801 flights made. This was an impressive record for those days, when the service had only sixty-three pilots and one hundred eighty enlisted men to keep the fifty airplanes serviceable and in the air.

COAST GUARD AVIATION IN WORLD WAR II

On 1 November 1941 the president put the coast guard under the operational control of the navy. Some months before, the number of both planes and personnel rapidly expanded. The efforts of coast guard air stations were primarily directed to antisubmarine patrols in 1942, 1943, and early 1944. Daily inshore patrols protected the important harbors and approaches to the sea, while offshore patrols covered other sections of our nation's coasts. Many enemy submarines were detected and several were attacked, damaged, or disabled. From the time of our entrance into the war until 30 June 1943, coast guard aircraft made sixty-one bombing attacks on enemy submarines. On 1 August 1942 coast guard Chief Aviation Pilot Henry C. White, flying a Grumman J4F-1 Amphibian, sank the German submarine U-166 in the Gulf of Mexico, about one hundred miles south of Houma, Louisiana. Action reports record numerous attacks that drove enemy submarines from coastal areas.

In October 1941 the Greenland Patrol was inaugurated by Captain Edward H. Smith, USCG, to operate as part of Task Force 24 of the U.S. Atlantic Fleet. Several months prior to the Pearl Harbor attack, coast guard cutter-based planes were operating in the Greenland area, carrying out antisubmarine and coastal patrols and making heroic rescues.

In the summer of 1943, to expand these efforts, the coast guard was directed to organize a special patrol squadron to be attached to the Greenland Patrol. Designated Patrol Bombing Squadron 6 and flying the rugged PBY-5A Catalina patrol bomber, this squadron became the most colorful of all World War II coast guard aviation units. Based in Greenland, it provided air coverage for ship convoys plying the North Atlantic between the United States and Britain, carried out antisubmarine patrols, surveyed ice conditions for shipping, made hundreds of dramatic rescues, and carried mail and supplies to remote regions. In addition to covering the vast Greenland area, the squadron's air operations ranged south to Newfoundland, west to the Hudson and Baffin bay regions, and east to Iceland. In 1943 it assisted coast guard cutters in locating, capturing, and destroying a Nazi weather station in northeast Greenland.

The coast guard was a leader in helicopter development and training. On 19 November 1943 the coast guard air station in Brooklyn, New York, was designated a helicopter training base. Here pilots and mechanics of various services and countries were trained. By February 1945 the school had trained 102 helicopter pilots—seventy-two coast guard, six navy, five army air force, twelve British, four Canadian, and three civilian pilots. A deck landing platform was installed in the air station to represent a rolling ship deck, and the coast guard cutter COBB was assigned to give actual landings at sea under varying weather conditions.

Besides the helicopter training program, the coast guard had many other important training assignments. A preflight school was maintained at its Elizabeth City air station in North Carolina, as was a school for aviation machinist mates and pilots destined for Greenland duty with Patrol Bombing Squadron 6. The coast guard air station at St. Petersburg, Florida, trained Mexican pilots for antisubmarine warfare.

The Air-Sea Rescue Agency was established in February 1944, but actually a unit had been organized in December 1943 at the coast guard air station in San Diego, California, as a result of the increasing number of airplane crashes in that region and to rescue fliers being forced down at sea. Investigation by the navy and coast guard showed that rescue equipment was adequate but rescue agencies were not well coordinated and the system of disseminating information was inefficient. The San Diego air station was the first air-sea rescue unit to be organized and put into operation in the United States. All surface craft, blimps, and planes, together with other rescue equipment used by the army, navy, marine corps, or other marine agencies, became a part of the new rescue organization. Actually, the coast guard, which had been organizing toward this end for some time, was the logical administrator for this new agency.

Coordination of army, navy, and coast guard rescue activities was controlled at joint operation centers, while actual rescues were the responsibility of each regional air-sea task unit, headed by the CO of the coast guard air station within that region. The national organization was headed by the commandant of the coast guard and an advisory board of representatives from other military services.

Besides rescue operations, air-sea-rescue services trained personnel engaged in search-and-rescue duties, and the coast guard carried on experimental work in the possibility of using blimps in air-sea rescue work. Actual rescues were made from raft to blimp, and a small number of officers and men were sent to school for training in blimp operations.

Other work of the rescue service included the organization of a special parachute unit, trained at the Forest Service's parachute jumpers school in Montana. Another unique air-sea-rescue training program was at the Port Angeles coast guard air station, where special air-land-rescue ski squads were trained and assisted by the national ski patrol to render service in snow-covered mountain areas.

The outstanding work of coast guard aviation during World War II remained little known to the general public. Its activities went far beyond the capabilities envisioned for it during peacetime. Statistics available for fiscal years 1943, 1944, and 1945 list 109,996,887 miles searched; time in flight, 285,054 hours; 1,388 persons rescued and assisted; 1,084 vessels and airplanes assisted; 515 medical cases transported; and 115,170 flights of all categories. In the performance of those exacting obligations, the coast guard exceeded even its own high standard of excellence and devotion to duty.

THE POST–WORLD WAR II YEARS

In January 1946 control of coast guard operations again returned to the Treasury Department. Congress, recognizing the need for an expanding coast guard aviation role, asked the treasury to determine its future aviation needs. The task was somewhat difficult. Many of the coast guard's missions were under review and yet undefined. Questions asked were how extensive America's territorial waters would be in the future and, given the trends in air traffic over the oceans, how far out search-and-rescue attempts would be undertaken. To accomplish its missions, the coast guard would require long-, medium-, and short-range aircraft.

The 1960s saw the emergence of the coast guard cutter/helicopter team as an effective part of the service's search and rescue, law enforcement, and military missions. Many new cutters in the medium- and high-endurance class, all with landing pads in their design, were commissioned. These new teams developed advanced techniques for helicopter utilization, and the helicopter gradually replaced the flying boat, which had long been associated with the coast guard.

As the coast guard entered the 1970s, expanding responsibilities brought new missions for aviation personnel. In 1976 the implementation of the 200-mile fishery conservation zone off U.S. shores

The big letters *USCG* have been a welcome sight to many a seaman in distress. Here crewmen of an HH-52 helicopter prepare to lower a pump to a sinking vessel. (U.S. Coast Guard)

expanded the role of the coast guard's fixed-wing aircraft and thereby expanded the service's responsibilities. The tremendous distances involved in the enforcement of the 200-mile zone created an additional need for high-speed aircraft capable of covering hundreds of square miles in a relatively short period of time. The new HU-25A Guardian meets these needs.

The role of aircraft continued to grow as a result of new coast guard responsibilities in environmental protection. In the mid-1970s an aerial oil-pollution surveillance system was incorporated in the Grumman HU-16Es and later added to the Lockheed HC-130 aircraft. This sophisticated electronic detection system can be used under all weather conditions, day or night, to detect the presence of oil on the ocean's surface.

The heavy-lift, long-range Lockheed C-130 Hercules allows the coast guard to search large areas at considerable distances offshore and to carry emergency supplies for air drop. (U.S. Coast Guard)

AIRCRAFT OPERATIONS OF THE 1980s

The tasks of coast guard aircraft have continued to grow. Of these, search-and-rescue operations receive the most public attention. Search and rescue is performed primarily for recreational boaters, whose numbers have increased a hundredfold since World War II. Annually, approximately half of the coast guard's operational flying hours are devoted to assisting them. Aircraft are used in the Arctic, Antarctic, and domestic icebreaking program. They evaluate ice conditions and recommend routes through the ice with the aid of surveillance planes.

The coast guard has primary responsibility for marine environmental protection, and its aircraft play a large part in this program. They conduct offshore and port patrols to see that laws are obeyed and to monitor environmental disasters. The coast guard has the most formidable and efficient pollution strike team in the world and can be ready at a moment's notice to fly anywhere if needed.

In navigable waters the coast guard enforces federal laws concerning marine safety, customs and revenue, immigration, fishing, and quarantine. Long-range aircraft operating from air stations and helicopters flying from cutters are an important part of the team used to interdict drug smuggling. They are also used for surveillance "to see and be seen."

The coast guard's worldwide radio-navigation aid, LORAN C, receives logistic and emergency support from long-range aircraft. The short-range aids-to-navigation program is also supported by aircraft. They search for missing aids, check signals, and give emergency service and construction and logistic support.

The International Ice Patrol, established as a result of the Titanic disaster, is supported by aircraft tracking and reporting dangerous icebergs that drift into the world's shipping lanes. Coast guard aircraft, on request, support many U.S. and Canadian government agencies.

The daily operations of coast guard aircraft may be controlled at a variety of administrative levels, depending on the urgency and

A duty officer takes an alert call at the coast guard air station in Miami. (U.S. Navy)

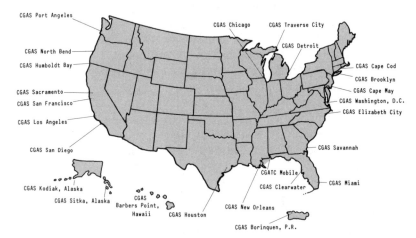

Fig. 15-1. Coast Guard Air Stations

nature of the assignment. Generally, a coast guard district assigns search-and-rescue missions to the air station. The air station then assigns standby aircraft to the mission. Operations may also be directed at the area and headquarters levels.

COAST GUARD AIR STATIONS

Twenty-seven coast guard air stations are strategically located to render the greatest service and support to all of the coast guard's missions.

COAST GUARD AVIATION TRAINING CENTER

The U.S. Coast Guard Aviation Training Center, in Mobile, Alabama, is a place of action. Although it carries out the functions normally assigned to coast guard air stations, its primary mission is training. It has the most sophisticated training syllabus in the coast guard. While most coast guard aviators graduate from the navy flight training program in Pensacola, Florida, recognized as the best in the world, many are unfamiliar with the missions, techniques, and machinery of the coast guard. To solve this problem and establish a standard flight training program throughout the coast guard, the aviation training center was commissioned in 1965.

The center's transitional courses for newly designated aviators provide standardized training in procedures and techniques peculiar to coast guard rotary-wing operations. Qualification courses are also given for new fixed-wing pilots and for fixed-wing pilots

A Sikorsky HH-3F helicopter from the coast guard air station in Cape Cod hoists a victim of a boating accident aboard for quick transfer to a local hospital. (Jess C. Barrow)

making the transition to instrument-qualified pilots and copilots of rotary-wing aircraft.

The training center also gives comprehensive refresher training. Each year pilots from throughout the coast guard come to the center to review and update their knowledge of air-traffic-control procedures, coast guard air operations, in-flight aircraft emergencies, and other aviation subjects.

Cadets who want a career in coast guard aviation and who have completed a two-year curriculum at the U.S. Coast Guard Academy in New London, Connecticut, visit the center for a two-week summer course. The cadets receive approximately fifty-three hours of instruction and ten hours of flight time. This course acquaints future officers with the resources and capabilities of aviation and stimulates interest in aviation careers.

One of the most interesting and exciting units of the coast guard is trained at the center. The shipboard helicopter operations division trains helicopter pilots and aircrews to be deployed aboard coast guard cutters and icebreakers for ice-breaking and polar missions, including scientific research, supply of remote stations, air logistics support, ice observations, polar search and rescue, law enforcement, and treaty enforcement.

Personnel are trained at the center for many other missions as well. The center takes great pride in the fact that its rotary-wing simulation program is one of the finest in the world.

PRINCIPAL COAST GUARD AIRCRAFT TODAY

HC-130 Hercules

The HC-130 Hercules is an all-weather, high-performance, four-engine, turboprop, long-range aircraft. As a search aircraft, it can proceed 1,200 nautical miles at 25,000 to 30,000 feet at 300 knots, let down to search altitude, search for two and a half hours at optimum search speed with two engines shut down, restart the idle engines, and return to base with reserve fuel. The HC-130 is an extremely versatile aircraft. It is capable of transporting up to ninety-two passengers, 35,000 pounds of cargo, and large quantities of rescue-survival equipment either for aerial delivery or transportation to the scene of a disaster. It was designed to land on short, unprepared airfields. It can be converted into a tanker configuration capable of refueling both fixed-wing and rotary-wing aircraft in the air.

HU-25A Guardian

The HU-25A Guardian is a medium-range surveillance aircraft. The operating parameters of this plane are broad based, however, and allow for a wide range of mission performance. It can operate from sea level to an altitude of 42,000 feet and fly at a maximum true air speed of up to 380 knots at sea level and 470 knots at 40,000 feet. The plane can proceed to the search area at high altitude and dash speed. It is an exceptionally stable platform for visual search in the 180- to 200-knot range and for the delivery of pumps or rafts at 130 knots. The Guardian is equipped with state-of-the-art avionics, built by Rockwell International, Collins Avionics Group. This system increases the probability of locating the search object once on scene, since it provides off-shore navigation, automatic search-pattern generation, redundant altitude control, and monitoring and communication capabilities on all marine aeronautical and government frequencies.

The HU-25A Guardian can proceed quickly to a vessel in distress with equipment to deal with the emergency. (Jess C. Barrow)

The coast guard's newest helicopter is the HH-65A Dolphin. It will be used for search and rescue, law enforcement, marine environmental protection, and marine science activities. (U.S. Coast Guard)

HH-3F Pelican

The HH-3F Pelican is an amphibious, twin-turbine, medium-range helicopter. As a search-and-rescue aircraft, it can proceed 300 nautical miles at 125 knots, hover for twenty minutes or land on the water, pick up eight survivors, and return to base with reserve fuel. Alternatively, the HH-3F can proceed 220 miles from base, search for two and one half hours, and return to base with reserve fuel. The HH-3F is equipped with the latest communication and navigation systems. This aircraft has a useful payload of 8,224 pounds and can be configured to carry passengers, cargo, stretchers, and small vehicles or boats. The HH-3F has a cargo ramp to facilitate loading.

HH-52A Sea-Guard

The HH-52A Sea-Guard is an amphibious, single-turbine, short-range helicopter. As a search-and-rescue aircraft, it can, with escort, proceed 150 nautical miles offshore at 90 knots, hover for twenty minutes or land on water, pick up four survivors, and return to base with reserve fuel.

HH-65A Dolphin

In 1983 the coast guard acquired a new generation of helicopter, the HH-65A Dolphin. The primary mission of the HH-65A is search and rescue; however, due to its versatility, this helicopter can be used to support many coast guard missions. The HH-65A has a 150-nautical-mile range at a cruising speed of 140 knots while transporting passengers. It can accommodate five passengers, including the crew. The maximum range without passengers is 380 nautical miles. It can carry 1,350 pounds in a cargo sling.

THE COAST GUARD AVIATORS' COMMISSIONING PROGRAM

The coast guard has approximately eight hundred aviation billets. Sixty percent of these are for helicopter-trained pilots and forty percent for fixed-wing-trained pilots. The service adds approximately sixty pilots a year to compensate for attrition. Pilots are obtained from three sources. Two-thirds are officers on their first entry-level assignment and are graduates of the Coast Guard Academy. They must have passed the academic qualification test, flight-aptitude rating, and flight physical in order to apply for postgraduate aviation training.

The second source of coast guard aviators is from the service's officer candidate school classes.* The coast guard graduates between four and five classes of officer candidates a year. On average, two to four candidates from each class are selected for flight training. Selection is made on the basis of qualification.

The third source of coast guard aviators is the direct commission aviator program. Two boards a year select prior military pilots (grade 0-2 or higher) who are college graduates and graduates of a U.S. military flight training program. Candidates must be a cer-

---

*The term *coast guard aviator* has been used throughout this chapter. However, all U.S. Coast Guard officers who complete the U.S. Navy's flight training program are designated naval aviators.

tain age and have a minimum of military flight time on their record. Those selected are given three-year contracts as lieutenants (j.g.) in an active-duty reserve status. Many of these individuals eventually join the regular coast guard. The direct commission program is used to augment the first two sources, and therefore the number of individuals selected varies greatly from year to year.

All coast guard officers selected under the first two programs receive their flight training at the naval air station in Pensacola. Upon completion they are designated naval aviators and receive wings of gold.

Appendixes

A

Navy Wings

The origin of distinctive insignia
for naval aviators is somewhat
obscure, but the idea was un-
doubtedly influenced by the fact
that U.S. Army aviators had been
wearing special badges since 1913.
The first correspondence on the
subject appears to have been a
letter from the CNO to the Bureau of Navigation, dated 19 July
1917, which forwarded a sugges-
tion from the G. F. Hemsley Co.
for aviators' caps and collar or-
naments, with the comment that
they were not especially wanted,
but since foreign countries and the
U.S. Army had adopted aviation
insignia, naval aviators should also
be given "some form of mark or

badge to indicate their qualifications, in order that they have standing with other aviation services." The letter enclosed a design to give an idea of what should be adopted.

The aviator wing insignia was officially adopted on 7 September 1917, when the secretary of the navy approved change 12 to uniform regulations. The pertinent portion read: "A Naval Aviator's device, a winged foul anchor with the letters *U.S.*, is hereby adopted to be worn by qualified Naval Aviators. This device will be issued by the Bureau of Navigation to Officers and Men of the Navy and Marine Corps who qualify as Naval Aviators, and will be worn on the left breast." Before any such wings were issued, the design was modified by another change of 12 October 1917, which deleted the letters *U.S.*

Correspondence continued with a number of firms concerning the design and production of the insignia. The first wings, made by Bailey, Banks & Biddle of Philadelphia, were received by the navy in December 1917 and issued early the following year. The design was described in more detail in uniform regulations approved on 20 September 1922.

One early design for naval aviator's wings may have looked like this.

Naval Aviator

Since then, the original design has undergone only a few minor changes.

NFOs can trace their origins back to Rear Admiral William A. Moffett, the first chief of the Bureau of Aeronautics and the first naval aviation observer. Observer's wings consisted of a single erect foul anchor encompassed by a circle to which wings were affixed. NFOs adopted new wings in 1968. These featured a shield in the center over two crossed foul anchors.

Today the old observer's wings are worn by flight meteorologists and certain marine corps officers who are graduates of the Marine Corps Naval Aviation Observer School at New River, North Carolina.

Navy navigator wings were replaced in 1968 by NFO wings for all qualified navy personnel. Marine corps enlisted and warrant-officer navigators still wear navigator wings to signify their qualifications.

Aircrew wings first came on the scene during World War II. These wings were designated combat-aircrew insignia in 1958 and may still be worn by marine corps personnel who qualified by virtue of service in Vietnam or Lebanon.

Navy personnel who have met

Naval Flight Officer

Naval Aviation Observer

Marine Aerial Navigator

Combat Aircrew

the requirements now wear aircrewmen wings, the centerpiece of which consists of a circle encompassing a single fouled anchor and the letters *AC*.

Aircrewman

Enlisted aircrewmen who have met the requirements may also wear the wings of an aviation warfare specialist. Recipients must be petty officers with at least two years in a squadron or on shipboard duty in an aviation assignment. They must also have demonstrated certain professional skills to qualify.

Aviation Warfare Specialist

Flight surgeons' wings have been changed several times over the years. Navy physicians become qualified to wear these wings after completing training at the Naval Aerospace Medical Institute in Pensacola.

Flight Surgeon

Wings may also be worn by a relatively small group within the aviation community designated naval aviation experimental psychologists and naval aviation physiologists. The centerpiece of these wings is the gold oak leaf of the medical service corps.

Naval Aviation Experimental
Psychologist/Physiologist

A special variation of naval aviators' wings belongs to the naval astronaut. A shooting star is superimposed on the traditional wings to symbolize the astronaut's special environment.

The most recent addition to the

Naval Astronaut

family of navy wings is awarded to those who meet the special qualifications of a naval aviation supply officer. The first set of these wings was presented by the DCNO (air warfare) in 1984.

Naval Aviation Supply Officer

B

Naval Aircraft

Prior to 1962 the letter-number systems used to identify aircraft and their functions were left to individual services to devise. Since that time, however, all U.S. military aircraft have been designated under a common system.

BASIC MISSION SYMBOL (LETTER)

The basic mission of an aircraft is identified by one of the following letters:

- A — Attack
- B — Bomber
- C — Cargo/transport
- E — Special electronic installation
- F — Fighter
- H — Helicopter
- K — Tanker
- O — Observation
- P — Patrol
- R — Reconnaissance
- S — Antisubmarine

T — Training
U — Utility
X — Research

The letter *V* may be used immediately after one of the above letters to signify that the aircraft has a vertical takeoff and landing (VTOL) capability, a short takeoff and landing (STOL) capability, or both (V/STOL). The Harrier attack aircraft (AV-8) and the Bronco observation aircraft (OV-10) are examples.

MISSION MODIFICATION SYMBOL (LETTER)

When an aircraft is modified so that its original mission is changed, a mission modification letter precedes the basic mission letter.

A — Attack
C — Cargo
D — Director (drone control)
E — Special electronic installation
H — Search and rescue
K — Tanker
L — Cold weather
M — Missile carrier
Q — Drone
R — Reconnaissance
S — Antisubmarine
T — Training
U — Utility
V — Staff
W — Weather

The TA-4 Skyhawk trainer, the EP-3 Orion electronic reconnaissance aircraft, and the SH-3 Sea King helicopter are examples of how mission modification symbols are used.

DESIGN SYMBOL (NUMBER)

The mission letter symbol or symbols are followed by a dash and then by the design symbol (number) of the aircraft. For example, the design symbol for the Hercules C-130 aircraft is 130. All of these aircraft, regardless of their mission or design modification, can be identified by that number as aircraft of that design type.

Thus, the EC-130, KC-130, and LC-130 can all be easily identified as variations of the original C-130 design.

DESIGN MODIFICATION SYMBOL (LETTER)

If an aircraft design itself has been modified, the design symbol (number) is followed immediately by a design modification symbol (letter). This letter simply indicates the number of times a particular aircraft design has been modified. The F-14D Tomcat fighter, for example, is the fourth modification of the original F-14 design.

SPECIAL-USE SYMBOL (LETTER)

Special-use symbols are infrequently seen preceding basic-mission or mission-modification symbols to indicate the special status of a particular aircraft.

G — Permanently grounded (for ground training)
J — Special test, temporary (when tests are complete, the aircraft will be restored to its original function)
N — Special test, permanent
X — Experimental
Y — Prototype
Z — In early stages of planning or development

CURRENT INVENTORY

Naval aircraft in the current inventory are listed below. The basic designation, name, and manufacturer of each aircraft is followed by the designations of all variations of that type.

Attack

A-3 Skywarrior (McDonnell Douglas)
EA-3B
KA-3B
ERA-3B
TA-3B
VA-3B

A-4 Skyhawk (McDonnell Douglas)
A-4E
A-4F
EA-4F
TA-4F
TA-4J
A-4M
OA-4M

A-6 Intruder (Grumman)
EA-6A
EA-6B
KA-6D
A-6E

A-7 Corsair II (LTV)
A-7B
A-7C
TA-7C
A-7E

F/A-8 Hornet (McDonnell Douglas)
F/A-18A
TF/A-18A

AV-8 Harrier V/STOL
(McDonnell Douglas/British Aerospace)
AV-8A
TAV-8A
AV-8B
AV-8C

Fighters

F-4 Phantom (McDonnell Douglas)
QF-4B
RF-4B
F-4J
EF-4J
F-4N
F-4S

F-5 Tiger II (Northrop)
F-5E
F-5F

F-8 Crusader (LTV)
RF-8G

F-14 Tomcat (Grumman)
F-14A
F-14D

Utility

OV-10 Bronco STOL (Rockwell
International)
OV-10A
OV-10D

**Patrol
Antisubmarine Warfare**

P-3 Orion (Lockheed)
P-3A
EP-3A
RP-3A
VP-3A
P-3B
EP-3B
P-3C
RP-3D
EP-3E

**Carrier-Based
Antisubmarine Warfare**

S-3 Viking (Lockheed)
S-3A
US-3A
S-3B

Airborne Early Warning

E-2 Hawkeye (Grumman)
TE-2A
TE-2B
E-2B
E-2C

Cargo Transport

C-1 Trader (Grumman)
C-1A

C-2 Greyhound (Grumman)
C-2A

C-9 Skytrain II (McDonnell Douglas)
C-9B

C-12 Super King Air (Beech)
UC-12B

C-130 Hercules (Lockheed)
C-130F
KC-130F
LC-130F
EC-130G
EC-130Q
KC-130R
LC-130R
KC-130T

C-131 Samaritan (Convair)
C-131F
C-131G
C-131H

Trainers

T-2 Buckeye (Rockwell International)
T-2B
T-2C

C-4 Academe/Gulfstream (Grumman)
TC-4C

T-34 Mentor (Beech)
T-34B
T-34C

T-39 Sabreliner (Rockwell International)
T-39D
CT-39E
CT-39G

T-44 Pegasus (Beech)
T-44A

T-45
(McDonnell Douglas/British Aerospace)
T-45A
T-47 Citation (Cessna)
T-47A

Helicopters
UH-1 Iroquois/Huey (Bell-Textron)
UH-1E
HH-1K
TH-1L
UH-1L
UH-1M
UH-1N
VH-1N

AH-1 Sea Cobra (Bell-Textron)
AH-1G
AH-1J
AH-1T

H-2 Seasprite-LAMPS Mk 1 (Kaman)
SH-2F

H-46 Sea Knight (Boeing Vertol)
HH-46A
UH-46A
CH-46D
UH-46D
CH-46E
CH-46F

H-3 Sea King (Sikorsky)
HH-3A
SH-3A
UH-3A
VH-3A
SH-3D
VH-3D
SH-3G
SH-3H

H-53 Sea Stallion (Sikorsky)
CH-53A
CH-53D
RH-53D

H-53E Super Stallion (Sikorsky)
CH-53E
MH-53E

H-57 Sea Ranger (Bell-Textron)
TH-57A
TH-57C

SH-60 Seahawk LAMPS Mk 3 (Sikorsky)
SH-60B

C

Navy and Marine Corps Squadrons

REGULAR NAVY SQUADRONS—Atlantic

Squadron	*Nickname*
ATKRON	
VA-12	Clinchers
VA-15	Valions
VA-34	Blue Blasters
VA-35	Black Panthers
VA-37	Bulls
VA-42	Green Pawns
VA-45	Blackbirds
VA-46	Clansmen
VA-55	Warhorses
VA-65	Tigers
VA-66	Waldos
VA-72	Blue Hawks
VA-75	Sunday Punchers
VA-81	Sunliners
VA-82	Marauders
VA-83	Rampagers

VA-85	Black Falcons
VA-86	Sidewinders
VA-87	Golden Warriors
VA-105	Gunslingers
VA-174	Hell Razors
VA-176	Thunderbolts

FITRON

VF-11	Red Rippers
VF-14	Tophatters
VF-31	Tomcatters
VF-32	Swordsmen
VF-33	Tarsiers
VF-41	Black Aces
VF-43	Challengers
VF-74	Bedevilers
VF-84	Jolly Rogers
VF-101	Grim Reapers
VF-102	Diamondbacks
VF-103	Sluggers
VF-142	Ghostriders
VF-143	Pukin Dogs
VF-171	Aces

STRIKFITRON

VFA-106	Gladiators
VFA-131	—
VFA-132	—
VFA-136	—
VFA-137	—

TACELRON

| VAQ-33 | Firebirds |
| VAQ-34 | Electric Horsemen |

CARAEWRON

| VAW-120 | Greyhawks |

VAW-121	Bluetails
VAW-122	Steeljaws
VAW-123	Screwtops
VAW-124	Bear Aces
VAW-125	Torchbearers
VAW-126	Seahawks
VAW-127	Seabats

AIRANTISUBRON

VS-22	Checkmates
VS-24	Scouts
VS-28	Hukkers
VS-30	Diamond Cutters
VS-31	Top Cats
VS-32	Maulers

HELANTISUBRON

HS-1	Seahorses
HS-3	Tridents
HS-5	Night Dippers
HS-7	Shamrocks
HS-9	Sea Griffins
HS-11	Dragon Slayers
HS-15	Red Lions

PATRON

VP-5	Mad Foxes
VP-8	Tigers
VP-10	Red Lancers
VP-11	Proud Pegasus
VP-16	Eagles
VP-23	Sea Hawks
VP-24	Batmen
VP-26	Tridents
VP-30	Pros Nest
VP-44	Golden Pelicans
VP-45	Pelicans

| VP-49 | Woodpeckers |
| VP-56 | Dragons |

FAIRECONRON

| VQ-2 | — |
| VQ-4 | Shadows |

FLECOMPRON

VC-6	Big Orange
VC-8	Redtails
VC-10	Challengers

FLELOGSUPPRON

| VR-24 | Lifting Eagles |
| VRC-40 | Codfish Airlines |

AIRFERRON

| VRF-31 | Storkliners |

AIRTEVRON

| VX-1 | Pioneers |

OCEANDEVRON

| VXN-8 | World Travellers |

HELANTISUBRON LIGHT

HSL-30	Neptune's Horsemen
HSL-32	Invaders
HSL-34	Professionals
HSL-36	Lamplighters
HSL-43	Battlecats

HELSUPPRON

HC-4	Black Stallions
HC-6	Chargers
HC-16	Bullfrogs

HELMINERON

HM-12	Sea Dragons
HM-14	Vanguard
HM-16	Seahawks

ACTIVE NAVY SQUADRONS—Pacific

ATKRON

VA-22	Fighting Redcocks
VA-27	Royal Maces
VA-52	Knightriders
VA-56	Champions
VA-93	Ravens
VA-94	Mighty Shrikes
VA-95	Green Lizards
VA-97	Warhawks
VA-115	Eagles
VA-122	Flying Eagles
VA-127	Cylons
VA-128	Golden Intruders
VA-145	Swordsmen
VA-146	Blue Diamonds
VA-147	Argonauts
VA-165	Boomers
VA-192	Golden Dragons
VA-195	Dam Busters
VA-196	Milestones

FITRON

VF-1	Wolfpack
VF-2	Bounty Hunters
VF-21	Freelancers
VF-24	Renegades
VF-51	Screaming Eagles
VF-111	Sundowners
VF-114	Aardvarks
VF-124	Gunfighters

VF-126	Bandits
VF-151	Vigilantes
VF-154	Black Knights
VF-161	Chargers
VF-211	Checkmates
VF-213	Black Lions

STRIKFITRON

VFA-25	Fist of the Fleet
VFA-113	Stingers
VFA-125	Rough Raiders

TACELRON

VAQ-129	Red Devils
VAQ-130	Zappers
VAQ-131	Lancers
VAQ-132	Scorpions
VAQ-133	Wizards
VAQ-134	Garudas
VAQ-135	Black Ravens
VAQ-136	Gauntlets
VAQ-137	Rooks
VAQ-138	Yellow Jackets
VAQ-139	Cougars

CARAEWRON

VAW-110	Firebirds
VAW-112	Golden Hawks
VAW-113	Black Eagles
VAW-114	Hormel Hawgs
VAW-115	Liberty
VAW-116	Sun Kings
VAW-117	Wallbangers

AIRANTISUBRON

VS-21	Fighting Redtails
VS-29	Dragonflies

VS-33	Screwbirds
VS-37	Sawbucks
VS-38	Red Griffins
VS-41	Shamrocks

HELANTISUBRON

HS-2	Golden Falcons
HS-4	Black Knights
HS-6	Indians
HS-8	Eight Ballers
HS-10	Taskmasters
HS-12	Wyverns

PATRON

VP-1	Screaming Eagles
VP-4	Skinny Dragons
VP-6	Blue Sharks
VP-9	Golden Eagles
VP-17	White Lightnings
VP-19	Big Red
VP-22	Blue Geese
VP-31	Black Lightnings
VP-40	Fighting Marlins
VP-46	Grey Knights
VP-47	Golden Swordsmen
VP-48	Boomerangers
VP-50	Blue Dragons

FAIRECONRON

| VQ-1 | World Watchers |
| VQ-3 | TACOMOPAC |

FLECOMPRON

| VC-1 | Blue Alii |
| VC-5 | Checkertails |

FLELOGSUPPRON

VRC-30	Truckin Traders
VRC-50	Foo Dogs

AIRTEVRON

VX-4	Evaluators
VX-5	Vampires

ANTARCTICDEVRON

VXE-6	Puckered Penguins

HELANTISUBRON

HSL-31	Archangels
HSL-33	Sea Snakes
HSL-35	Magicians
HSL-37	Easy Riders
HSL-41	Seahawks

HELSUPPRON

HC-1	Fleet Angels
HC-3	Packrats
HC-5	Providers
HC-11	Gunbearers

TRAINING SQUADRONS

Squadron	Nickname	Location
TRARON		
VT-2	Doer Birds	NAS Whiting, FL
VT-3	Red Knights	NAS Whiting, FL
VT-4	Rubber Ducks	NAS Pensacola, FL
VT-6	—	NAS Whiting, FL
VT-7	Eagles	NAS Meridian, MS
VT-9	Tigers	NAS Meridian, MS
VT-10	Cosmic Cats	NAS Pensacola, FL
VT-19	Fighting Frogs	NAS Meridian, MS
VT-21	Red Hawks	NAS Kingsville, TX

VT-22	King Eagles	NAS Kingsville, TX
VT-23	Professionals	NAS Kingsville, TX
VT-24	Bobcats	NAS Chase Field, TX
VT-25	Cougars	NAS Chase Field, TX
VT-26	Tigers	NAS Chase Field, TX
VT-27	Boomers	NAS Corpus Christi, TX
VT-28	Rangers	NAS Corpus Christi, TX
VT-31	Wise Owls	NAS Corpus Christi, TX
VT-86	Sabre Hawks	NAS Pensacola, FL

HELTRARON

| HT-8 | — | NAS Whiting, FL |
| HT-18 | — | NAS Whiting, FL |

ATKRON

VA-203	Blue Dolphins	NAS Ceal Field, FL
VA-204	River Rattlers	NAS New Orleans, LA
VA-205	—	NAS Atlanta, GA
VA-304	Firebirds	NAS Alameda, CA
VA-305	Lobos	NAS Pt. Mugu, CA

FITRON

VF-201	Hunters	NAS Dallas, TX
VF-202	Superheats	NAS Dallas, TX
VF-301	Devil's Disciples	NAS Miramar, CA
VF-302	Fighting Stallions	NAS Miramar, CA

STRIKFITRON

| VFA-303 | Golden Hawks | NAS Alameda, CA |

LIGHTPHOTRON

| VFP-206 | Hawkeyes | NAF Washington, D.C. |

AERREFRON

| VAK-208 | Jockeys | NAS Alameda, CA |
| VAK-308 | Griffins | NAS Alameda, CA |

TACELRON

VAQ-209	Star Warriors	NAS Norfolk, VA
VAQ-309	—	NAS Whidbey Island, WA

CARAEWRON

VAW-78	Fighting Escargots	NAS Norfolk, VA
VAW-88	Cottonpickers	NAS Miramar, CA

HELANTISUBRON

HS-74	Minutemen	NAS South Weymouth, MA
HS-75	Emerald Knights	NAS Willow Grove, PA
HS-84	—	NAS North Island, CA
HS-85	Golden Gators	NAS Alameda, CA

PATRON

VP-60	Cobras	NAS Glenview, IL
VP-62	Broadarrows	NAS Jacksonville, FL
VP-64	Condors	NAS Willow Grove, PA
VP-65	Tridents	NAS Pt. Mugu, CA
VP-66	—	NAS Willow Grove, CA
VP-67	Golden Hawks	NAS Memphis, TN
VP-68	Blackhawks	NAS Patuxent River, MD
VP-69	Totems	NAS Whidbey Island, WA
VP-90	Lions	NAS Glenview, IL
VP-91	Stingers	NAS Moffett Field, CA
VP-92	Minutemen	NAS South Weymouth, MA
VP-93	Executioners	NAF Detroit, MI
VP-94	Crawfishers	NAS New Orleans, LA

FLECOMPRON

VC-12	Omars	NAS Oceana, VA
VC-13	—	NAS Pt. Mugu, CA

FLELOGSUPPRON

VR-46	Peach Airlines	NAS Atlanta, GA
VR-48	—	NAF Washington, D.C.
VR-51	—	NAS Glenview, IL

VR-52	Taskmasters	NAS Willow Grove, PA
VR-55	Bicentennial Minutemen	NAS Alameda, CA
VR-56	—	NAS Norfolk, VA
VR-57	—	NAS North Island, CA
VR-58	—	NAS Jacksonville, FL
VR-59	Express	NAS Dallas, TX
VR-60	—	NAS Memphis, TN
VR-61	Islanders	NAS Whidbey Island, WA

HELATKRON LIGHT

HAL-4	Redwolves	NAS Norfolk, VA
HAL-5	Blue Hawks	NAS Pt. Mugu, CA

HELSUPPRON

HC-9	Protectors	NAS North Island, CA

HELANTISUBRON LIGHT

HSL-74	Demon Elves	NAS South Weymouth, MA
HSL-84	—	NAS North Island, CA

MARINE AIRCRAFT SQUADRONS

Squadron	*Nickname*
Attack	
VMA-211	Wake Island Avengers
VMA-214	Blacksheep
VMA-223	Bulldogs
VMA-231	Spades
VMA-311	Tomcats
VMA-331	Bumblebees
VMA-513	Nightmares
VMA-542	Flying Tigers

All-Weather Attack

VMA (AW)-121	Green Knights
VMA (AW)-224	Bingers
VMA (AW)-242	Bats

VMA (AW)-332	Tophatters
VMA (AW)-533	Hawks

Fighter/Attack

VMFA-115	Silver Eagles
VMFA-122	Crusaders
VMFA-212	Lancers
VMFA-232	Red Devils
VMFA-235	Death Angels
VMFA-251	Thunderbolts
VMFA-312	Checkerboards
VMFA-314	Black Knights
VMFA-323	Death Rattlers
VMFA-333	Shamrocks
VMFA-451	Warlords
VMFA-531	Gray Ghosts

Tactical Electronic Warfare

VMAQ-2	Playboys

Tactical Recon

VMFP-3	Eyes and Ears of the Wing

Refueler Transport

VMGR-152	—
VMGR-252	—
VMGR-352	—

Observation

VMO-1	—
VMO-2	—

Attack and Fighter/Attack Training

VMAT-102	—
VMAT-203	—
VMAT (AW)-202	—
VMFAT-101	—

Attack Helicopter

HMA-169	—
HMA-269	Sea Cobras
HMA-369	—

Light Helicopter

HML-167	—
HML-267	—
HML-269	—
HML-367	—

Medium Helicopter

HMM-161	—
HMM-162	Dough Boys
HMM-163	Ridge Runners
HMM-164	—
HMM-165	White Knights
HMM-261	Bulls
HMM-262	Tigers
HMM-263	Thunder Chickens
HMM-264	Black Knights
HMM-265	—
HMM-268	—
HMM-365	—

Heavy Helicopter

HMH-361	Flying Tigers
HMH-362	Ugly Angels
HMH-363	Lucky Red Lions
HMH-461	Sea Stallions
HMH-462	Heavy Haulers
HMH-463	Pineapples
HMH-464	—
HMH-465	—

Helicopter Training

HMT-204	—

HMT-301   —
HMT-303   —

Marine Helicopter Squadron One
HMX-1   —

MARINE AIR RESERVE SQUADRONS

Squadrons	Nickname	Location
VMA-124	—	NAS Memphis, TN
VMA-131	Snakes	NAS Willow Grove, PA
VMA-133	—	NAS Alameda, CA
VMA-134	Hawks	MCAS El Toro, CA
VMA-142	Gators	NAS Jacksonville, FL
VMA-322	Fighting Cocks	NAS South Weymouth, MA
VMFA-112	Cowboys	NAS Dallas, Texas
VMFA-321	Flight Barons	NAF Washington, D.C.
VMAQ-4	—	NAS Whidbey Island, WA
VMGR-234	—	NAS Glenview, IL
VMO-4	—	CGAS Detroit, MI
HMA-773	—	NAS Atlanta, GA
HML-771	Hummers	NAS South Weymouth, MA
HML-776	—	NAS Glenview, IL
HMM-764	—	MCAS Santa Ana, CA
HMM-767	—	NAS New Orleans, LA
HMM-774	Honkers	NAS Norfolk, VA
HMH-772	—	NAS Willow Grove
HMH-772 Det. A	—	NAS Alameda, CA
HMH-772 Det. B	—	NAS Dallas, TX
HMH-777	Heavy Haulers	NAS Dallas, TX

D

U.S. Naval Aviation Ships

Type of Ship	Mission
CV/CVN—multipurpose aircraft carrier	Supports and operates aircraft to engage in attacks on targets at sea and ashore
ARVA—aircraft repair ship aircraft	Provides logistic support and repair facilities to aviation units operating at unequipped air bases
ARVE—aircraft repair ship engine	Provides logistic support and repair facilities to aviation units operating at unequipped air bases
AVM—guided-missile ship	Provides a mobile guided-missile launching platform
AVS—aviation supply ship	Transports and issues aviation supplies
LPH—amphibious assault ship	Transports and lands troops, equipment, and supplies by helicopter
AVT—auxiliary aircraft transport	Transports and delivers aircraft and their components
AKV—cargo ship and aircraft ferry (aviation)	Transports replacement aircraft, general aviation stores, and general cargo

Type of Ship	*Mission*
LHA—amphibious-assault ship (general purpose)	Embarks, deploys, and lands elements of a marine landing force in an assault by helicopters, landing craft, and amphibian vehicles
AVT—auxiliary aircraft-landing training ship	Trains and qualifies naval aviators in carrier landings

MULTIPURPOSE CARRIERS

Midway (CV 41)
  Length 977' 2"
  Beam 243'
  Displacement 64,222 tons
  Flt.-deck length 977' 2"
  Flt.-deck width 258' 6"
  Home port   Yokosuka, Japan
  Air wing   CVW-5
Coral Sea (CV 43)
  Length 978'
  Beam 231'
  Displacement 62,700 tons
  Flt.-deck length 978'
  Flt.-deck width 236'
  Home port   Norfolk, Virginia
  Air wing   CVW-13
Forrestal (CV 59)
  Length 1046'
  Beam 250' 8"
  Displacement 78,200 tons
  Flt.-deck length 1040'
  Flt.-deck width 252' 8"
  Home port   Mayport, Florida
  Air wing   CVW-6
Saratoga (CV 60)
  Length 1046'
  Beam 250' 8"
  Displacement 78,200 tons

Flt.-deck length 1040'
Flt.-deck width 252' 8"
Home port   Mayport, Florida
Air wing   CVW-17

Ranger (CV 61)
Length 1047' 6"
Beam 255'
Displacement 79,200 tons
Flt.-deck length 1047' 6"
Flt.-deck width 249' 11"
Home port   San Diego, California
Air wing   CVW-2

Independence (CV 62)
Length 1047' 6"
Beam 255'
Displacement 79,200 tons
Flt.-deck length 1047' 6"
Flt.-deck width 249' 11"
Home port   Norfolk, Virginia

Kitty Hawk (CV 63)
Length 1047' 6"
Beam 251' 8"
Displacement 80,000 tons
Flt.-deck length 1047' 6"
Flt.-deck width 250'
Home port   San Diego, California
Air wing   CVW-9

Constellation (CV 64)
Length 1047' 6"
Beam 251' 8"
Displacement 80,000 tons
Flt.-deck length 1047' 6"
Flt.-deck width 250'
Home port   San Diego, California
Air wing   CVW-14

Enterprise (CVN 65)
Length 1101'

Beam 253' 11"
Displacement 89,400 tons
Flt.-deck length 1101' 6"
Flt.-deck width 248' 3"
Home port   Alameda, California
Air wing   CVW-11

America (CV 66)
  Length 1047' 6"
  Beam 252'
  Displacement 78,600 tons
  Flt.-deck length 1047' 6"
  Flt.-deck width 267'
  Home port   Norfolk, Virginia
  Air wing   CVW-1

John F. Kennedy (CV 67)
  Length 1051'
  Beam 270' 6"
  Displacement 81,086 tons
  Flt.-deck length 1051'
  Flt.-deck width 267'
  Home port   Norfolk, Virginia
  Air wing   CVW-3

Nimitz (CVN 68)
  Length 1092'
  Beam 252' 8"
  Displacement 91,685 tons
  Flt.-deck length 1092'
  Flt.-deck width 250' 8"
  Home port   Norfolk, Virginia
  Air wing   CVW-8

Dwight D. Eisenhower (CVN 69)
  Length 1092'
  Beam 252' 8"
  Displacement 91,505 tons
  Flt.-deck length 1092'
  Flt.-deck width 250' 8"
  Home port   Norfolk, Virginia
  Air wing   CVW-7

Carl Vinson (CVN 70)
  Length 1092'
  Beam 252' 8"
  Displacement 91,400 tons
  Flt.-deck length 1092'
  Flt.-deck width 250' 8"
  Home port   Alameda, California
  Air wing   CVW-15

TRAINING CARRIERS

Lexington (AVT 16)
  Length 890' 3"
  Beam 171' 3"
  Displacement 44,200 tons
  Flt.-deck length 884' 7"
  Flt.-deck width 162' 9"
  Home port   Pensacola, Florida
  Assignment Training

CARRIERS UNDER CONSTRUCTION OR APPROVED

Theodore Roosevelt (CVN 71)
  Keel laid   31 October 1981
  Christened   27 October 1984
  Commissioning October 1986 (planned)
Abraham Lincoln (CVN 72)
  Keel laid   3 November 1984
  Contract delivery   December 1989
George Washington (CVN 73)
  Contract delivery   December 1991

AMPHIBIOUS ASSAULT SHIPS, GENERAL PURPOSE (LHA)

Tarawa (LHA 1)
  Length 320'
  Beam 106'
  Displacement 39,400 tons
  Fleet assignment   Pacific
Saipan (LHA 2)
  Length 820'

Beam 106'
Displacement 39,400 tons
Fleet assignment   Atlantic
Belleau Wood (LHA 3)
　Length 820'
　Beam 106'
　Displacement 38,900 tons
　Fleet assignment   Pacific
Nassau (LHA 4)
　Length 820'
　Beam 106'
　Displacement 38,900 tons
　Fleet assignment   Atlantic
Pelelieu (LHA 5)
　Length 820'
　Beam 106'
　Displacement 38,900 tons
　Fleet assignment   Pacific

AMPHIBIOUS ASSAULT SHIPS (LPH)

Iwo Jima (LPH 2)
　Length 602'
　Beam 84'
　Displacement 18,042 tons
　Fleet assignment   Atlantic
Okinawa (LPH 3)
　Length 598'
　Beam 84'
　Displacement 18,154 tons
　Fleet assignment   Pacific
Guadalcanal (LPH 7)
　Length 598'
　Beam 84'
　Displacement 18,000 tons
　Fleet assignment   Atlantic
Guam (LPH 9)
　Length 598'
　Beam 84'

Displacement 18,300 tons
Fleet assignment   Atlantic
Tripoli (LPH 10)
  Length 592'
  Beam 84'
  Displacement 18,515 tons
  Fleet assignment   Pacific
New Orleans (LPH 11)
  Length 592'
  Beam 84'
  Displacement 18,241 tons
  Fleet assignment   Pacific
Inchon (LPH 12)
  Length 598'
  Beam 84'
  Displacement 18,825 tons
  Fleet assignment   Atlantic

AMPHIBIOUS ASSAULT SHIPS UNDER CONSTRUCTION OR APPROVED
Wasp (LHD 1)
  Length 844'
  Beam 106'
  Displacement 40,500 tons
  Construction begun 1984
  Scheduled for delivery 1989

GUIDED-MISSILE SHIPS

Norton Sound (AVM 1)
  Length 540'
  Beam 71'
  Displacement 15,170 tons
  Fleet assignment   Pacific

RESERVE AIRCRAFT CARRIERS

Hornet (CVS 12)
Bennington (CVS 20)
Bon Homme Richard (CV 31)
Oriskany (CV 34)

E

U.S. Naval Air Stations, Air Facilities, and Other Airfields

ATLANTIC FLEET SUPPORT

NAS Bermuda
NAS Brunswick, Maine
NAS Cecil Field, Florida
ALF Fentress (aux. Oceana)
NAS Guantanamo Bay, Cuba
NAS Jacksonville, Florida
NAVSTA Keflavik, Iceland
NAS Key West, Florida
NAF Lajes, Azores
NAF Mayport, Florida
NAF Mildenhall, England
NAVSUPACT Naples, Italy
NAS Norfolk, Virginia
NAS Oceana, Virginia
NAVSTA Roosevelt Roads, Puerto Rico
NAVSTA Rota, Spain
NAS Sigonella, Scilly
OLF Whitehouse (outlying Cecil)

PACIFIC FLEET SUPPORT

NAS Adak, Alaska
NAS Agana, Guam
NAS Alameda, California
NAF Atsugi, Japan
NAS Barbers Point, Hawaii
OLF Coupeville (outlying Whidbey)
ALF Crows Landing (aux. Moffett)
NAS Cubi Point, Philippines
NAF Diego Garcia, Indian Ocean
NAF El Centro, California
NAS Fallon, Nevada
ALF Ford Island (aux. Barbers Point)
OLF Imperial Beach (outlying North Island)
NAF Kadena, Okinawa, Japan
NAS Lemoore, California
NAF Midway Island
NAS Miramar
NAF Misawa, Japan
NAS Moffett Field, California
NAS North Island, California
ALF San Clemente (outlying North Island)
NAS Whidbey Island, Washington

RESEARCH DEVELOPMENT AND TEST

PMRF Barking Sands, Hawaii (facility of Point Mugu)
NWC China Lake, California
NAEC Lakehurst, New Jersey
NATC/NAS Patuxent River, Maryland
PMTC/NAS Point Mugu, California
OLF San Nicholas Island (outlying Point Mugu)
NADC Warminster, Pennsylvania

EDUCATION AND TRAINING

OLF Alpha (outlying Meridian)
OLF Barin (outlying Whiting)
OLF Bravo (outlying Meridian)

OLF Brewton (outlying Whiting)
OLF Bronson (outlying Pensacola)
ALF Cabaniss (aux. Corpus Christi)
NAS Chase Field, Texas
OLF Choctaw (outlying Pensacola)
NAS Corpus Christi, Texas
ALF Goliad (aux. Chase Field)
OLF Harold (outlying Whiting)
OLF Holley (outlying Whiting)
NAS Kingsville, Texas
NAS Memphis, Tennessee
NAS Meridian, Mississippi
OLF Middleton (outlying Whiting)
ALF Orange Grove (outlying Kingsville)
OLF Pace (outlying Whiting)
NAS Pensacola, Florida
OLF Santa Rosa (outlying Whiting)
OLF Saufley Field (outlying Whiting)
OLF Silver Hill (outlying Whiting)
ALF Site 6 (aux. Pensacola)
ALF Site 8 (aux. Whiting)
OLF Spencer (outlying Whiting)
OLF Summerdale (outlying Whiting)
ALF Waldron (aux. Corpus Christi)
NAS Whiting Field, Florida
OLF Wolf (outlying Whiting)

NAVAL AIR RESERVE STATIONS AND FACILITIES

NAS Atlanta, Georgia
NAS Dallas, Texas
NAF Detroit, Michigan
NAS Glenview, Illinois
NAS New Orleans, Louisiana
NAS South Weymouth, Massachusetts
NAS Willow Grove, Pennsylvania
NAF Washington, D.C.

MARINE CORPS AIR STATIONS AND FACILITIES

MCAS Beaufort, South Carolina
MCALF Bogue Field, South Carolina
MCAF Camp Pendleton, California
MCAS Cherry Point, North Carolina
MCAS El Toro, California
MCAS(H) Futema, Okinawa
MCAS Iwakuni, Japan
MCAS Kaneohe Bay, Hawaii
MCAS(H) New River, North Carolina
MCAF Quantico, Virginia
MCAS(H) Tustin, California
MCAS Yuma, Arizona

COAST GUARD AIR STATIONS

Astoria, Oregon
Barbers Point, Hawaii
Borinquen, Puerto Rico
Brooklyn, New York
Cape Cod, Massachusetts
Cape May, New Jersey
Chicago, Illinois
Clearwater, Florida
Corpus Christi, Texas
Detroit, Michigan
Elizabeth City, North Carolina
Houston, Texas
Humboldt Bay, California
Kodiak, Alaska
Los Angeles, California
Miami, Florida
New Orleans, Louisiana
North Bend, Oregon
Port Angeles, Washington
Sacramento, California
San Diego, California

San Francisco, California
Savannah, Georgia
Sitka, Alaska
Traverse City, Michigan
Washington, D.C.
Coast Guard Aviation Training Center (CGATC), Mobile, Alabama

Active Navy and Marine Corps Airfields Named for Individuals

Airfield Name	Location	Named for
Albert Mitchell Field	NAS Adak, AL	Ensign Albert E. Mitchell, USN
Alvin Callender Field	NAS New Orleans, LA	Captain Alvin A. Callender, RFC
Appollo Soucek Field	NAS Oceana, VA	Vice Admiral Appollo Soucek, USN
Armitage Field	NAVWPNCEN China Lake, CA	Lieutenant John W. Armitage, USN
Arthur W. Radford Field	NAS Cubi Point, Philippines	Admiral Arthur W. Radford, USN
Ault Field	NAS Whidbey Island, Oak Harbor, WA	Commander William B. Ault, USN
Brewer Field	NAS Agana, Guam	Commander Charles W. Brewer, USN
Bristol Field	NAVSTA Argentia, Newfoundland	Admiral Arthur L. Bristol, USN
Cabaniss Field	NALF Cabaniss, Corpus Christi, TX	Commander Robert W. Cabaniss, USN
Cecil Field	NAS Cecil Field, FL	Commander Henry B. Cecil, USN
Chambers Field	NAS Norfolk, VA	Captain Washington I. Chambers, USN
Chase Field	NAS Chase Field, Beesville, TX	Lieutenant Commander Nathan B. Chase
Cunningham Field	MCAS Cherry Point, NC	Lieutenant Colonel Alfred A. Cunningham, USMC
Forrest Sherman Field	NAS Pensacola, FL	Admiral Forrest Sherman, USN
Frederick C. Sherman Field	NALF San Clemente, CA	Admiral Frederick C. Sherman, USN

Airfield Name	Location	Named for
Halsey Field	NAS North Island, San Diego, CA	Admiral William F. Halsey, USN
Henderson Field	NAF Midway Island, HI	Major Lofton R. Henderson, USMC
Hensley Field	NAS Dallas, TX	Colonel William N. Hensley, USMC
John Rodgers Field	NAS Barbers Point, HI	Commander John Rodgers, USN
John Towers Field	NAS Jacksonville, FL	Admiral John H. Towers, USN
Joseph Mason Reeves Field	NAS Lemoore, CA	Admiral Joseph M. Reeves, USN
Maxfield Field	Lakehurst, NJ	Commander Louis H. Maxfield, USN
McCain Field	NAS Meridian, MI	Admiral John S. McCain, USN
McCalla Field	NAS Guantanamo, Cuba	Captain Bowman H. McCalla, USN
McCutcheon Field	MCAS New River, NC	General Keith B. McCutcheon, USMC
Merritt Field	MCAS Beaufort, SC	Major General Lewis G. Merritt, USMC
Mitcher Field	NAS Miramar, CA	Admiral Marc A. Mitcher, USN
Moffett Field	NAS Moffett Field, CA	Rear Admiral William A. Moffett, USN
Nimitz Field	NAS Alameda, CA	Fleet Admiral Chester W. Nimitz, USN
Ofstie Field	NAVSTA Roosevelt Roads, Puerto Rico	Vice Admiral Ralph A. Ofstie, USN
Ream Field	OLF Imperial Beach, CA	Major William R. Ream, USN
Saufley Field	OLF Saufley Field, Pensacola, FL	Lieutenant (j.g.) Richard C. Saufley, USN
Shea Field	NAS South Weymouth, MA	Lieutenant Commander John J. Shea, USN
Turner Field	MCAF Quantico, VA	Colonel Thomas C. Turner, USMC
Van Voorhis Field	NAS Fallon, NE	Commander Bruce Van Voorhis, USN

Airfield Name	Location	Named for
Waldron Field	NALF Waldron, Corpus Christi, TX	Lieutenant John C. Waldron, USN
Webster Field	OLF Webster, Priest's MD	Captain Walter W. Webster, USN
Whiting Field	NAS Whiting Field, Milton, FL	Captain Kenneth Whiting, USN
Williams Field	McMurdo Sound, Antarctica	Richard Williams

F

Congressional Medal Winners

The Medal of Honor is this country's highest award for valor. It originated during the Civil War for those who exhibited courage "above and beyond the call of duty." It was first authorized for the navy and marine corps on 21 December 1861 and for the army and voluntary forces on 12 July 1862.

The medal is awarded in the name of Congress and for this reason is often called the Congressional Medal of Honor. President Theodore Roosevelt, on 20 September 1905, ordered the award to be made ceremoniously and the recipient to be in Washington, D.C., for the presentation, which would be made by the president, as commander in chief, or by a representative of the president.

The medal is made in three different designs, one for the sea services (navy, marine corps, and coast guard), one for the army, and one for the air force. The navy medal is bronze, suspended by an anchor from a bright blue ribbon worn about the neck. The ribbon contains a cluster of thirteen white stars representing the original thirteen states. The medal itself is a five-pointed star, each ray of which contains sprays of laurel and oak tipped with a trefoil. Minerva, personifying the Union, stands in the center, surrounded by thirty-four stars representing the thirty-four states that existed in 1861. She holds in her left hand an axe bound in staves of wood, the ancient Roman symbol of authority. The shield in her right hand repulses the serpents held by the crouching figure of Discord. On the reverse side of the medal the recipient's name and the date and place of the act of valor are engraved.

The following list of names includes naval aviators, a naval aviation observer, and others whose actions are associated with naval aviation. Some of those listed became part of the naval aviation community after the action that won them their Medals of Honor. Ranks given are those held at the times of the action.

Antrim, Richard N., Lieutenant, USN—Action in behalf of fellow prisoners while POW, April 1942

Bauer, Harold W., Lieutenant Colonel, USMC—Air combat, South Pacific, 28 Sep.–3 Oct. 1942 (posthumous)

Bennett, Floyd, Chief Warrant Officer, USN—Pilot on first flight over North Pole, 9 May 1926

Boyington, Gregory, Major, USMC—Air combat, Central Solomon Islands, 12 Sep. 1943–3 Jan. 1944

Byrd, Richard E., Lieutenant Commander, USN—Commander of first flight over North Pole, 9 May 1926

Clausen, Raymond M., Private First Class, USMC—Repeated rescues by helicopter of men under fire, South Vietnam, 30 Jan. 1970

Commiskey, Henry A., Second Lieutenant, USMC—Led ground attack on strong enemy position, Korea, 20 Sep. 1950

Corry, William M., Lieutenant Commander, USN—Attempted rescue of pilot from burning aircraft, 2 Oct. 1920 (posthumous)

De Blanc, Jefferson J., Captain, USMC—Air combat, Solomon Islands, 31 Jan. 1943

Edson, Merritt A., Colonel, USMC—Led ground combat action in defense of Henderson Field, Guadalcanal, 13–14 Sep. 1942

Elrod, Henry T., Captain, USMC—Air and ground combat in defense of Wake Island, 8–23 Dec. 1941 (posthumous)

Estocin, Michael J., Lieutenant Commander, USN—Air combat, North Vietnam, 20, 26 April 1967 (posthumous)

Finn, John W., Chief Petty Officer, USN—Action under fire during attack on NAS Kaneohe Bay, Hawaii, 7 Dec. 1941

Fleming, Richard E., Captain, USMC—Leader of dive-bombing attack during Battle of Midway, 4–6 June 1942 (posthumous)

Foss, Joseph J., Captain, USMC—Air combat in defense of Guadalcanal, 9 Oct.–19 Nov. 1942

Gary, Donald A., Lieutenant (j.g.), USN—Repeated rescues of trapped men aboard USS *Franklin* (CV 13) following severe combat damage, 19 Mar. 1945

Galer, Robert E., Major, USMC—Air combat, South Pacific, Aug.–Sep. 1942

Gordon, Nathan G., Lieutenant, USN—Repeated air rescues of men in the water under fire, New Ireland Island, 15 Feb. 1944

Hall, William E., Lieutenant (j.g.), USN—Determined attacks on enemy carrier, Battle of the Coral Sea, 7–8 May 1942

Hammann, Charles H., Ensign, USNRF—Rescue of fellow pilot under fire during raid on Pola, Austria, 21 Aug. 1918

Hanson, Robert M., First Lieutenant, USMC—Air combat in the Solomon Islands and at New Britain, 24 June 1942 (posthumous)

Hudner, Thomas J., Jr.—Attempted rescue of squadron mate downed behind enemy lines, Korea, 4 Dec. 1950

Hutchins, Carlton B., Lieutenant, USN—Remained at controls of aircraft following mid-air collision to allow crew to escape, 2 Feb. 1938 (posthumous)

Koelsch, John K., Lieutenant (j.g.), USN—Attempted rescue by helicopter under fire, Korea, 3 July 1951 (posthumous)

Lassen, Clyde E., Lieutenant (j.g.), USN—Night helicopter rescue under fire, North Vietnam, 19 June 1968

McCampbell, David, Commander, USN—Air combat during battles of Philippine Sea and Leyte Gulf, June and Oct. 1944

McDonnell, Edward O., Ensign, USN—Established signal station ashore and maintained communications under fire, Veracruz, 21–22 Apr. 1914

McGunigal, Patrick, Petty Officer First Class, USN—Rescued Kite balloon pilot entangled underwater, 17 Sep. 1917

Moffett, William A., Commander, USN—Action in command of a ship at Veracruz, 21–22 Apr. 1914

O'Callahan, Joseph T., Lieutenant Commander, CHC., USN—Inspiration, leadership and repeated rescues aboard USS *Franklin* (CV 13) following severe combat damage, 19 Mar. 1945

O'Hare, Edward H., Lieutenant, USN—Air combat in defense of carrier off Rabaul, 20 Feb. 1942

Ormsbee, Francis E., Jr., Chief Machinist's Mate, USN—Rescued crewman and attempted to rescue pilot in crash of seaplane, Pensacola Bay, 25 Sep. 1918

Pless, Stephen W., Captain, USMC—Helicopter rescue under fire, Vietnam, 19 Aug. 1967

Powers, John J., Lieutenant, USN—Determined attacks on enemy ships during Battle of the Coral Sea, 4–8 May 1942 (posthumous)

Ricketts, Milton E., Lieutenant, USN—Led damage-control party aboard USS *Yorktown* during Battle of the Coral Sea, 8 May 1942 (posthumous)

Robinson, Robert G., GSGT., USMC—Air combat as gunner in Europe, 8, 14 Oct. 1918

Schilt, Christian F., First Lieutenant, USMC—Air evacuation of wounded under fire, Nicaragua, 6–8 Jan. 1928

Smith, John L., Major, USMC—Air combat in defense of Guadalcanal, 21 Aug.–15 Sep. 1942

Stockdale, James B., Captain, USN—Action in behalf of fellow prisoners while POW, Vietnam, 4 Sep. 1969

Swett, James E., First Lieutenant, USMC—Air combat, Solomon Islands, 7 April 1943

Talbot, Ralph, Second Lieutenant, USMC—Air combat, Europe, 8, 14 Oct. 1918

Van Voorhis, Bruce A., Lieutenant Commander, USN—Determined bomber attack, Battle of the Solomon Islands, 6 July 1943 (posthumous)

Walsh, Kenneth A., First Lieutenant, USMC—Air combat, Solomon Islands, 15, 30 Aug. 1943

CONGRESSIONAL SPACE MEDAL OF HONOR WINNERS

The Congressional Space Medal of Honor was approved by Congress in 1969 for presentation "to any astronaut who in the performance of his duties has distinguished himself by exceptionally meritorious efforts and contributions to the welfare of the nation and of mankind." The following is a list of naval aviators or former naval aviators who received this award.

Armstrong, Neil A.—Gemini 8, 1966, and Apollo 11, 1969, when Armstrong became the first person to walk on the moon
Conrad, Charles, Jr.—Four spaceflights from 1965–73, culminating in the first manned Skylab mission
Glenn, John H., Jr.—Mercury, 1962, first American to orbit the earth
Shepard, Alan B., Jr.—Mercury, 1961, first American in space; commander of Apollo 14, third lunar landing mission in 1971
Young, John W.—Commander of the first orbital flight of the Space Shuttle in April 1981; also a veteran of two Gemini and two Apollo flights

NC-4 CONGRESSIONAL MEDAL

Congress ordered seven of these medals struck in solid gold following the world's first successful transatlantic flight by the Navy Curtiss NC-4 in May 1919. One medal was presented to each of the six crewmen of the NC-4. The seventh medal was presented to the commander of the entire three-plane flight, whose aircraft was one of two that went down at sea and did not complete the crossing. Recipients of the NC-4 Congressional Medal are:

Breese, James L., Lieutenant, USN
Hinton, Walter, Lieutenant (j.g.), USN
Read, Albert C., Lieutenant Commander, USN
Rhoads, Eugene S., Chief Machinists Mate, USN
Rodd, Herbert C., Ensign, USN
Stone, Elmer F., Lieutenant, USCG
Towers, John H., Commander, USN

G

Naval Aviation Hall of Honor

On 10 July 1980 the names of the first twelve distinguished men to be enshrined in the Naval Aviation Hall of Honor were approved by the CNO. Most were early naval aviators, but the list also included the first naval aviation observer and two civilians. All were enshrined at a ceremony opening the hall at the Naval Aviation Museum in Pensacola on 6 November 1981.

Since that time others have been added to the list. All are carefully chosen by a special selection committee appointed by the CNO. This committee makes its selections based on the following criteria:

—Sustained superior performance in or for naval aviation
—Superior contributions in the technical or tactical development of naval aviation
—Unique and superior flight achievement in combat or noncombat flight operations

A bronze plaque commemorating the achievements and/or special contributions of each honoree is placed on display in the Hall of Honor following the ceremonies.

Naval Aviation Hall of Honor

MEMBERS OF THE NAVAL AVIATION HALL OF HONOR

Admiral John H. Towers, USN—Naval aviator no. 3; pioneer of naval aviation; officer in charge of the navy's first formal flying school at Pensacola; headed navy Curtiss NC flying boat project; first commander navy Seaplane Division 1; commanded NC-3 in transatlantic attempt; chief of the Bureau of Aeronautics during WW II buildup; flag rank leader throughout WW II

Eugene Burton Ely—Civilian pioneer of naval aviation; first take-off of an aircraft from a ship (USS *Birmingham*); made first landing-takeoff cycle to and from a ship (USS *Pennsylvania*)

Lieutenant Colonel Alfred A. Cunningham, USMC—Naval aviator no. 5, marine corps aviator no. 1; commanded first marine corps aviation squadron; CO of First Marine Aviation Force in Europe, WW I

Rear Admiral Richard E. Byrd, USN—Naval aviator no. 608; devoted more than 30 years to polar exploration

Commander Theodore G. Ellyson, USN—Naval aviator no. 1; pioneer of naval aviation; involved in earliest efforts to win acceptance of aviation in the navy

Glenn Hammond Curtiss—Civilian pioneer of naval aviation; built the navy's first aircraft (Triad A-1); trained the navy's first pilot; leading designer and manufacturer of early naval aircraft; codesigner of navy Curtiss NC-4

Vice Admiral Patrick N. L. Bellinger, USN—Naval aviator no. 8; pioneer of naval aviation; first naval aviator to come under hostile fire (Vera Cruz); commander of NC-1 in transatlantic attempt; flag rank leader of WW II

Rear Admiral William A. Moffett, USN—Naval aviation observer no. 1; first chief of the Bureau of Aeronautics; strong advocate of naval aviation between the wars

Rear Admiral Albert C. Read, USN—Naval aviator no. 24; pilot in command of NC-4 during the world's first flight across the Atlantic; flag rank leader of WW II

Lieutenant Commander Godfrey de C. Chevalier, USN—Naval aviator no. 7; commanded the Northern Bombing Group in France during WW I; first U.S. naval aviator to land aboard a carrier (USS *Langley*)

Captain Holden C. Richardson, USN—Naval aviator no. 13; early aircraft designer; supervised construction of the NC flying boats; served as pilot of NC-3 in transatlantic attempt

Warrant Officer Floyd Bennett, USN—Early naval aviation pilot; served as pilot for Lieutenant Commander Richard E. Byrd on first flight over the North Pole

General Roy M. Geiger, USMC—Naval aviator no. 49; marine corps aviator no. 5; pioneer of marine corps aviation; commanded First Air Wing at Guadalcanal and held the line against the Japanese onslaught

Glenn L. Martin—Civilian aviation pioneer and founder of the Glenn L. Martin Company; produced many aircraft for naval aviation and is especially known for his giant flying boats

Admiral Marc A. Mitscher, USN—Naval aviator no. 33; pioneer of naval aviation; pilot of NC-1 during transatlantic attempt;

outstanding commander of fast carrier task forces in the Pacific in WW II

Admiral Arthur W. Radford, USN—Naval aviator no. 2896; director of naval air training at the outset of WW II; carrier division commander in the Pacific; first naval officer to serve as chairman, Joint Chiefs of Staff; outspoken defender of naval aviation

Vice Admiral Charles E. Rosendahl, USN—Naval aviator no. 3174; pioneer of the navy's great rigid airships of the 1920s; survivor of the crash of the airship Shenandoah and commander of the airships Akron and Los Angeles; outspoken advocate of lighter-than-air operations in the navy

Commander Elmer F. Stone, USCG—Naval aviator no. 38; coast guard aviator no. 1; one of the pilots of the NC-4 which made the world's first transatlantic crossing by air; strong proponent of coast guard aviation

Captain Kenneth Whiting, USN—Naval aviator no. 16; commanded the First Aeronautical Detachment in Europe; developed U.S. base structure overseas during WW I; influential in the development of carrier operations aboard USS *Langley*

Leroy R. Grumman—Naval aviator no. 1216; cofounder of the Grumman Aircraft Engineering Corp.; manufacturer of naval aircraft including the Wildcat, Hellcat, and Avenger of WW II

Vice Admiral James H. Flatley, Jr., USN—Fighter ace, air group commander, aggressive combat leader of WW II; involved in the development of the fighter tactic known as the "Thach Weave" for combating Japanese Zeros

Admiral John S. Thach, USN—Fighter ace, renowned tactician and originator of the "Thach Weave" for engaging Japanese Zeros

H

Naval Aviation Awards

Award	Sponsor(s)	Presented	Selection Criteria
Ancient Albatross Award	U.S. Coast Guard	As appropriate	To the active-duty coast guard pilot who has held his wings for the longest period of time
Aviation Ground Officer of the Year Award (marine)	U.S. Marine Corps Aviation Association	Annually	To the top marine aviation ground officer
Aviation Supply Excellence Award (S-3 support)	Lockheed California	Annually	To the supply support activity (COMNAVAIRLANT, COMNAVAIRPAC) that demonstrates the most effective aviation supply support of S-3 type aircraft
Aviation Supply Excellence Award (P-3 support)	Lockheed California	Annually	To the supply support activity (COMNAVAIRLANT, COMNAVAIRPAC and CNAVRES) that demonstrates the most effective aviation supply support of the P-3 type aircraft
Aviator's Valor Award	American Legion	Annually	To an aviator for a conspicuous act of valor or courage performed during aerial flight, in or out of combat (nominee can be from any of the military services)
Battle Es	Chief of Naval Operations	Annually	To Atlantic and Pacific Fleet units that are outstanding in combat readiness
Bronze Hammer Award (self-help)	Chief of Naval Operations	Annually	To activities that have made the most progress in self-help improvements, using available resources

Award	Sponsor(s)	Presented	Selection Criteria
Commandants Efficiency Award	U.S. Marine Corps Aviation Association	Annually	To the marine corps squadron whose efficiency, readiness, and operational flexibility are the best
Arleigh Burke Fleet Trophy	Chief of Naval Operations	Annually	To the ship or squadron in the Atlantic and Pacific fleets demonstrating the greatest improvement in battle efficiency
Admiral Joseph P. Clifton Award	Litton Industries	Annually	To the most outstanding fighter squadron in the navy, based on operational performance and safety, administrative excellence, and effective aircraft maintenance
Edwin Francis Conway Trophy	Chief of Naval Operations	Annually	To the reserve naval air station, naval air facility, or naval air reserve unit judged to be the most effective in the performance of its primary mission
Chief of Naval Operations Safety Awards	Chief of Naval Operations	Annually	To the major commands that contribute the most toward readiness, high morale, and economy of operations through safety
Alfred A. Cunningham Award	Marine Corps Aviation Association	Annually	To the outstanding marine corps aviator of the year
Noel Davis Trophy	Chief of Naval Operations	Annually	To the top naval air reserve squadrons, based on excellence in mobilization readiness

Award	Sponsor	Frequency	Description
F. Trubee Davison Award	McDonnell Douglas Corp.	Annually	To the naval air reserve tailhook squadron that excels in combat readiness and carrier proficiency
Leonardo Da Vinci Golden Helix Award	Navy Helicopter Association Sikorsky Aircraft Co.	As appropriate	To the naval aviator serving on active duty with the U.S. Navy, Marine Corps, or Coast Guard who bears the earliest date of designation as a navy helicopter pilot
Admiral James H. Flatley, Jr., Award	Rockwell International	Annually	To aviation ships for superior operational readiness, an outstanding safety record, and significant contributions to the field of aviation safety
Major Joseph Foss Award	American Fighter Aces Association	Annually	To an American military student pilot for excellence in the air-combat maneuvering stage of flight training
Vice Admiral Robert Goldthwaite Award	Rockwell International	Annually	To the top training squadron in recognition of outstanding achievement in the training of naval aviators within the Naval Air Training Command
Grampaw Pettibone Award	Chief of Naval Operations and Mr. Paul H. Warner	Annually	To the individual or organization that has contributed most toward naval aviation safety awareness through articles, publications, or art
Gray Eagle Award	LTV Aerospace and Defense Co.	As appropriate	To the naval aviator on active duty with the earliest date of designation as a naval aviator

Award	Sponsor(s)	Presented	Selection Criteria
Gray Owl Award	Grumman Aerospace Corp.	As appropriate	To the NFO on active duty who has held that designation for the longest period
Robert M. Hanson Award	Marine Corps Aviation Association	Annually	To the top marine corps fighter squadron of the year
Helicopter Squadron of the Year Award (marine)	Marine Corps Aviation Association	Annually	To the top marine corps helicopter squadron of the year
David S. Ingalls Award	The Navy League	Annually	To the top instructor in the Naval Air Training Command based on flight safety, officer-like qualities, character, personality, and leadership
Captain Arnold J. Isbell Trophy	Lockheed-California Co.	Annually	To squadrons judged to be the best in air antisubmarine warfare during each 12-month competitive cycle
Thurston H. James Memorial Award	General Commandery of the Naval Order of the United States	Annually	To the outstanding graduate of the NFO program
LAMPS Ship Safety Award	Rockwell International	Annually	To a LAMPS-configured ship in recognition of an outstanding safety record in LAMPS operations
Lifelong Service Recognition Award	Navy Helicopter Association and the Bendix Corp.	Annually	To the individual who has, during his lifetime, contributed most significantly to vertical-lift development and the Navy Helicopter Assn.

Award	Sponsor	Frequency	Description
Rear Admiral Clarence Wade McClusky Award	LTV Aerospace and Defense Co.	Annually	To the best attack squadron in the navy, based on maintenance, operational safety, and administrative excellence
AVCM Donald M. Neal Award (also known as the Golden Wrench Award)	Lockheed	Annually	To the best antisubmarine and patrol squadron, based on excellence in aircraft maintenance (COMNAVAIRLANT, COMNAVAIRPAC & CNAVRES)
Outstanding Achievement Award (4 categories)	Association of Naval Aviation	Annually	To the individual or squadron in each of the following categories for outstanding achievement in naval aviation: helicopter operations maritime patrol aviation tactical aviation fleet support and special mission aviation
Admiral Arthur W. Radford Award	Grumman Aerospace Corp.	Annually	To the airborne-early-warning squadron that excels in operational readiness, safety, retention, and contributions to tactics and weapon systems development
Robert Guy Robinson Award	Marine Corps Aviation Association	Annually	To the outstanding marine corps NFO of the year
Admiral James S. Russell Naval Aviation Flight Safety Award	Order of Daedalians	Annually	To the major naval aviation command that has demonstrated excellence and effectiveness in aircraft accident prevention

Award	Sponsor(s)	Presented	Selection Criteria
Lawson H. M. Sanderson Award	Marine Corps Aviation Association	Annually	To the top marine corps attack squadron of the year
Secretary of the Navy Energy Conservation Awards	Secretary of the Navy	Annually	To the aviation commands that are outstanding in their efforts to reduce energy costs and promote responsible energy management
Sikorsky Helicopter Rescue Award	Sikorsky Aircraft	As appropriate	To search-and-rescue aircrewmen for skill and courage in helicopter rescue operations during life threatening situations
Silver Falcon Award	Association of Naval Aviation San Diego Squadron	As appropriate	To the senior marine corps reserve naval aviator
Silver Hawk Trophy	McDonnell Douglas Corp.	As appropriate	To the marine corps naval aviator on active duty whose date of designation precedes that of any other marine corps naval aviator
Villard C. Sledge Memorial Maintenance Awards	Chief of Naval Operations	Annually	To intermediate maintenance activities for outstanding accomplishment in the repair of jet engines
Admiral Jimmy Thach Award	Lockheed-California Co.	Annually	To the outstanding carrier-based antisubmarine, air or helicopter antisubmarine squadron, based on superior performance
Vice Admiral John H. Towers Award	Order of Daedalians	Annually	To the training squadron with the best mission-oriented safety record

Award	Sponsor(s)	Presented	Selection Criteria
V/STOL Squadron of the Year (marine)	Marine Corps Aviation Association	Annually	To the top marine corps V/STOL squadron of the year
Orville Wright Achievement Award	Order of Daedalians	Biannually	To outstanding graduates of U.S. military undergraduate training programs for flight proficiency, academic achievement, and officer-like qualities

I

Aces and MiG Killers*

FIGHTER ACES

In the seventy-year history of aerial combat, the subject of fighter aces and their victories has proven a subject of enduring interest. Navy and marine corps aviators, whether flying as tailhookers or from land bases, occupy a special niche in the history of American aces.

The following compilation of the naval service's top fighter pilots in each war is intended to update previous lists. Space limitations dictate that the WWII names are limited to those aces generally credited with ten or more kills. In the first column of figures are the commonly accepted victory scores, in this instance taken from R. F. Toliver's 1965 volume, *Fighter Aces*. The second column notes the tallies, where different, from Dr. Frank J. Olynyk, whose exhaustive research is acknowledged as the most authoritative to date. Ranks are those known in each case as the highest attained while active as a combat fighter pilot.

*Information in appendix I is reprinted from the winter 1983 issue of *The Hook*.

Any "aceologist" recognizes that a pilot's victory score is subject to many variables. Credits-versus-claims is only the starting point for discrepancies. Sometimes ground claims are lumped in with aerial victories, or a higher command issues totals at variance with lower echelons. Some contradictions simply defy explanation.

No U.S. air arm officially acknowledges the ace system, but the American Fighter Aces Association follows tradition in requiring five confirmed aerial victories as the minimum standard.

Regardless of his victory total, the successful pilot is responsible for achieving and maintaining air superiority with his squadron mates. This compilation reflects the records credited to the navy and marine corps' leading fighter pilots in the four air wars of this century.

As a note of interest to most tailhookers, a recent AFAA study confirms the F6F Hellcat produced more U.S. aces (three hundred) than any other aircraft of any service.

Material for this feature was compiled by J. R. "Bill" Bailey, Dr. Frank J. Olynyk, and Barrett Tillman, secretary of the American Fighter Aces Association.

UNITED STATES NAVY

World War I

	Toliver	Olynyk
LT David S. Ingalls		6

World War II

	Toliver	Olynyk
CDR David McCampbell (MOH)	34	
LT Cecil E. Harris	24	
LT Eugene A. Valencia	23	
LT Patrick D. Fleming	19	
LT Alexander Vraciu	19	
LT Cornelius N. Nooy	19	
LT Ira C. Kepford	17	16
LT Charles R. Stimpson	17	16
LTJG Douglas Baker	16	16.33
LT Arthur R. Hawkins	14	
LCDR Elbert S. McCuskey	14	13.5

LT John L. Wirth	14	13.5
LCDR George C. Duncan	13.5	
LCDR Roger W. Mehle	13.33	5.66
LT Daniel A. Carmichael	13	12
LT Dan R. Rehm	13	9
LT Roy W. Rushing	13	
LT John R. Strane	13	
LT Wendell V. Twelves	13	
LCDR Clement M. Craig	12	11.75
LT Leroy E. Harris	12	9.25
LCDR Roger R. Hedrick	12	
LCDR William E. Henry	12	9.5
LT William J. Masoner	12	
LCDR Hamilton McWhorter III	12	
LCDR Edward H. O'Hare (MOH)	12	7
LT James A. Shirley	12	12.5
LT George R. Carr	11.5	
CDR Fred E. Bakutis	11	7.5
CDR John T. Blackburn	11	
CDR William A. Dean	11	
LT James B. French	11	
LT Phillip L. Kirkwood	11	12
LT Charles M. Mallory	11	10
LTJG James V. Reber	11	
LCDR James F. Rigg	11	
LT Donald E. Runyon	11	
LT Richard E. Stambook	11	
LT Stanley W. Vejtasa	11	10.25
CDR Marshall U. Beebe	10.5	
LT Russell L. Reiserer	10.5	9
LT Armistead B. Smith	10.5	10
LT John C.C. Symmes	10.5	11
LT Albert O. Vorse	10.5	11.5
LT Robert E. Murray	10.33	
LT Robert H. Anderson	10	8
LT Carl A. Brown	10	10.5
LCDR Thaddeus T. Coleman	10	

LT Richard L. Cormier	10	8
LT Harris E. Mitchell	10	
LT T. Hamil Reidy	10	
LT Arthur Singer	10	
LT John M. Smith	10	
LT James S. Swope	10	9.66

Korea

LT Guy P. Bordelon	5

Vietnam

LT Randall Cunningham	5
LTJG Willie Driscoll (RIO)	5

UNITED STATES MARINE CORPS

World War II

MAJ Joseph J. Foss (MOH)	26	
1/LT Robert M. Hanson (MOH)	25	
MAJ Gregory Boyington (MOH)	22 + 3.5 with AVG	
CAPT Kenneth A. Walsh (MOH)	21	
CAPT Donald N. Aldrich	20	
MAJ John L. Smith (MOH)	19	
MAJ Marion E. Carl	18.5	
CAPT Wilbur J. Thomas	18.5	
CAPT James E. Swett (MOH)	16	15.5
CAPT Harold L. Spears	15	
MAJ Kenneth D. Frazier	14.5	13.5
CAPT Edward O. Shaw	14.5	
MAJ Archie G. Donahue	14	
MAJ James N. Cupp	13	12
MAJ Robert E. Galer (MOH)	13	14
1/LT William P. Marontate	13	
MAJ Loren D. Everton	12	
CAPT Harold E. Segal	12	
2/LT Eugene A. Trowbridge	12	6
CAPT William N. Snider	11.5	
CAPT Philip C. DeLong	11.17 (Plus 2 in Korea)	

LCOL Harold W. Bauer (MOH)	11	10
CAPT Donald H. Sapp	11	10
CAPT Jack E. Conger	10	
MAJ Herbert H. Long	10	

Korea

MAJ John F. Bolt	6	(Plus 6 in WWII)

AVG = American Volunteer Group (Flying Tigers)
MOH = Medal of Honor
RIO = Radar Intercept Officer

MiG KILLERS

The following is an unofficial listing of all confirmed and possible aerial victories scored by U.S. Navy and Marine Corps aircrews during the Vietnam War. Probables are indicated by (P) following the type of aircraft believed to be downed. The list was prepared from several official and unofficial sources. Wherever possible, the data has been confirmed by official sources, including many of the aircrews involved.

Date	Time	Mission	A/C	BuNo	Aircrew
09/04/65	0840	BarCap	F-4B	151403	LTJG Terence M. Murphy-KIA
					ENS Ronald J. Fegan-KIA
Remarks: *ChiCom; High altitude engagement; 602 MIA MiG guns?; Enemy reports downed by friendly					
17/06/65	1030	BarCap	F-4B	151488	CDR Louis Page 'Lou'
					LT John C. Smith, Jr.
17/06/65	1030	BarCap	F-4B	152219	LT Jack E.D. Batson, Jr. 'Dave'
					LCDR Robert B. Doremus
20/06/65	1835	ResCap	A-1H	137523	LT Charlie Hartman
Remarks: ½ kill					
20/06/65	1835	ResCap	A-1H	139768	LT Clinton B. Johnson 'Clint'
Remarks: ½ kill					
06/10/65	1040	BarCap	F4-B	150634	LCDR Dan MacIntyre
					LTJG Alan Johnson
Remarks: *Confirmed, not released					
12/06/66	1446	TarCap	F-8E	150924	CDR Harold L. Marr 'Hal'
12/06/66	1450	TarCap	F-8E	150924	CDR Harold L. Marr 'Hal'
21/06/66	1530	Photo/SAR	F-8E	150924	LTJG Phillip V. Vampatella 'Phil'
21/06/66	1535	ResCap	F-8E	150867	LT Eugene J. Chancy 'Gene'
13/07/66	1102	TarCap	F-4B	151500	LT William M. McGuigan 'Squeaky'
					LTJG Robert M. Fowler
09/10/66	0945	TarCap	F-8E	149159	CDR Richard M. Bellinger 'Dick'*
Remarks: *Deceased					

Date	Time	Mission	A/C	BuNo	Aircrew
09/10/66	1013	ResCap	A-1H	137543	LTJG William T. Patton
20/12/66	1832	Intercept	F-4B	153022	LT H. Dennis Wisely 'Denny'
					LTJG David L. Jordan
20/12/66	1841	Intercept	F-4B	153019	LT David McCrea 'Barrel'
					ENS David Nichols
24/04/67		TarCap	F-4B	153000	LT Charles E. Southwick 'Ev'
					ENS James W. Laing
Remarks: A/C lost this date; possible AAA					
24/04/67	1645	TarCap	F-4B	153037	LT H. Dennis Wisely 'Denny'
					LTJG Gareth L. Anderson
01/05/67	1244	Flak Supp	A-4C	148609	LCDR Theodore R. Swartz 'Ted'
01/05/67	1245	TarCap	F-8E	150303	LCDR Marshall O. Wright 'Moe'
19/05/67	1525	TarCap	F-8E	150348	CDR Paul H. Speer
19/05/67	1525	TarCap	F-8E	150661	LTJG Joseph M. Shea
Remarks: *'Herded' MiG with 20mm for position					
19/05/67	1525	Flak Supp	F-8C	146981	LCDR Bobby C. Lee
19/05/67	1525	TarCap	F-8C	147029	LT Phillip R. Wood
21/07/67	0830	TarCap	F-8C	147018	LCDR Marion H. Isaacks 'Red'
21/07/67	0830	TarCap	F-8C	146998	LTJG Phil Dempewolf
21/07/67	0830	TarCap	F-8C	146992	LCDR Robert L. Kirkwood
21/07/67	0830	Escort	F-8E	150859	LCDR Ray G. Hubbard 'Tim'
10/08/67	1232	TarCap	F-4B	152247	LT Guy H. Freeborn
					LTJG Robert J. Elliot
10/08/67	1232	TarCap	F-4B	150431	LCDR Robert C. Davis
					LCDR Gayle O. Elie 'Swede'
26/10/67	1300	MiGCap	F-4B	149411	LTJG Robert P. Hickey, Jr.
					LTJG Jeremy G. Morris 'Jerry'
30/10/67	1245	MiGCap	F-4B	150629	LCDR Eugene P. Lund 'Gino'
					LTJG James R. Borst, 'Bif'*
Remarks: A/C lost same day, own missile *Deceased					
14/12/67	1740	TarCap	F-8E	150879	LT Richard E. Wyman 'Dick'
09/05/68			F-4B	153036	MAJ John P. Hefferman, USAF 'Jack'
					LTJG Frank A. Schumacher
Remarks: *Confirmed, not released					
26/06/68	1810	MiGCap	F-8H	148710	CDR Lowell R. Meyers 'Moose'
09/07/68	0850	Escort	F-8E	150926	LCDR John B. Nichols, III
10/07/68	1600	MiGCap	F-4J	155553	LT Roy Cash, Jr.
					LT Joseph E. Kain, Jr.
29/07/68	1132	MiGCap	F-8E	150349	CDR Guy Cane
01/08/68	1340	MiGCap	F-8H	147916	LT Norman K. McCoy
19/09/68	1055	MiGCap	F-8C	146961	LT Anthony J. Nargi 'Tony'
28/03/70			F-4J	155875	LT Jerome E. Beaulier 'Jerry'
					LT Steven J. Barkley
19/01/72	1358	TarCap	F-4J	157267	LT Randall H. Cunningham 'Yank'
					LTJG William P. Driscoll 'Willie'
06/03/72	1325	ForCap	F-4B	153019	LT Garry L. Weigand 'Greyhound'
					LTJG William C. Freckelton 'Farkle'
06/05/72	1410	TarCap	F-4B	150456	LCDR Jerry B. Houston 'Devil'
					LT Kevin T. Moore
06/05/72	1825	MiGCap	F-4J	157249	LT Robert G. Hughes 'Bob'*
					LTJG Adolph J. Cruz 'Joe'

Remarks: *killed midair VF-126/121 A-4/F-4

Date	Time	Mission	A/C	BuNo	Aircrew
06/05/72	1825	MiGCap	F-4J	157245	LCDR Kenneth W. Pettigrew 'Viper'
					LTJG Michael J. McCabe 'Mike'
08/05/72	1005	MiGCap	F-4J	157267	LT Randall H. Cunningham 'Duke'
					LTJG William P. Driscoll 'Irish'
10/05/72	0958	TarCap	F-4J	157269	LT Curt Dose
					LCDR James McDevitt
10/05/72	1400	TarCap	F-4J	155769	LT Michael J. Connelly 'Matt'
					LT Thomas J.J. Blonski
10/05/72	1400	TarCap	F-4J	155769	LT Michael J. Connelly
					LT Thomas J.J. Blonski
10/05/72	1400	MiGCap	F-4B	151398	LT Kenneth L. Cannon 'Ragin Cajun'
					LT Roy A. Morris, Jr. 'Bud'
10/05/72	1400	TarCap	F-4J	155749	LT Steven C. Shoemaker 'Steve'
					LTJG Keith V. Crenshaw
10/05/72	1401	Flak Supp	F-4J	155800	LT Randall H. Cunningham
					LTJG William P. Driscoll
10/05/72	1403	Flak Supp	F-4J	155800	LT Randall H. Cunningham
					LTJG William P. Driscoll
10/05/72	1408	Flak Supp	F-4J	155800	LT Randall H. Cunningham
					LTJG William P. Driscoll

Remarks: A/C lost same day; SAM; crashed 2019/10640

Date	Time	Mission	A/C	BuNo	Aircrew
18/05/72	1730	MiGCap	F-4B	153068	LT Henry A. Bartholomay 'Bart'
					LT Oran R. Brown
18/05/72	1730	MiGCap	F-4B	153915	LT Patrick E. Arwood 'Pat'
					LT James M. Bell 'Mike'
23/05/72	1755	MiGCap	F-4B	153020	LCDR Ronald E. McKeown 'Mugs'
					LT John C. Ensch 'Jack'
23/05/72	1755	MiGCap	F-4B	153020	LCDR Ronald E. McKeown
					LT John C. Ensch
11/06/72	1045	MiGCap	F-4B	149473	CDR Foster S. Teague 'Tooter'
					LT Ralph M. Howell
11/06/72	1045	MiGCap	F-4B	149457	LT Winston W. Copeland 'Mad Dog'
					LT Donald R. Bouchoux 'Don'
21/06/72	1215	MiGCap	F-4J	157293	CDR Samuel C. Flynn, Jr. 'Sam'
					LT William H. John 'Bill'
10/08/72	2019	MiGCap	F-4J	157299	LCDR Robert E. Tucker, Jr. 'Gene'
					LTJG Stanley B. Edens 'Bruce'
11/09/72	1802	MiGCap	F-4J	155526	MAJ Lee T. Lasseter 'Bear'*
					CAPT John D. Cummings 'Li'l John'*

Remarks: A/C lost this date; SAM *Deceased

Date	Time	Mission	A/C	BuNo	Aircrew
28/12/72	1225	MiGCap	F-4J	155846	LTJG Scott H. Davis
					LTJG Geoffrey H. Ulrich 'Jeff'
12/01/73	1332	BarCap	F-4B	153045	LT Victor T. Kovaleski 'Vic'
					LTJG James A. Wise
		Escort	F-4D	667709	CAPT Doyle Baker, USMC
					1/LT John D. Ryan, Jr., USAF

Remarks: Exchange duty 432 TRW; Kill: 17 Dec 67; Location: Route Package 6A

		WX Reece	F-4E	670239	CAPT Lawrence G. Richard, USMC
					LCDR Michael J. Ettel, USN*

Remarks: Exchange duty 432 TRW; Kill: 12 Aug 72; *Died VF-43 A-4, Oceana, 1974

Squadron	Modex	Call Sign	CV	CVW	Weapon	Location	Kill
VF-96	NG 602	Showtime	CVA-61	CVW-9	AIM-7	1820/10830	MiG-17 (P)*
VF-21	NE 101	Sundown	CVA-41	CVW-2	AIM-7	2008/10515	MiG-17
VF-21	NE 102	Sundown	CVA-41	CVW-2	AIM-7	2008/10515	MiG-17
VA-25	NE 573	Canasta	CVA-41	CVW-2	20mm	2010/10525	MiG-17
VA-25	NE 577	Canasta	CVA-41	CVW-2	20mm	2010/10525	MiG-17
VF-151	NL 107	Switch Box	CVA-43	CVW-15	AIM-7D		MiG-17 (P)*
VF-211	NP	Nickel	CVA-19	CVW-21	AIM-9D	2120/10630	MiG-17
VF-211	NP	Nickel	CVA-19	CVW-21	AIM-9D	2120/10620	MiG-17 (P)
VF-211	NP 104	Nickel	CVA-19	CVW-21	AIM-9D	2133/10637	MiG-17D
VF-211	NP 101	Nickel	CVA-19	CVW-21	AIM-9D	2133/10637	MiG-17
VF-161	NL 216	Rock River	CVA-64	CVW-15	AIM-9D	2041/10555	MiG-17
VF-162	AH 210	Superheat	CVA-34	CVW-16	AIM-9	2132/10548	MiG-21
VA-176	AK 409	Papoose	CVS-11	CVW-10	20mm	2015/10522	MiG-17
VF-114	NH 215	Linfield	CVA-63	CVW-11	AIM-7E	1927/10558	An 2
VF-213	NH 110	Black Lion	CVA-63	CVW-11	AIM-7E	1927/10558	An 2
VF-114	NH 210	Linfield	CVA-63	CVW-11	AIM-9B	2123/10616	MiG-17
VF-114	NH 00	Linfield	CVA-63	CVW-11	AIM-9D	2123/10616	MiG-17
VA-76	NP 685	Sun Glass	CVA-31	CVW-21	Zuni	2121/10620	MiG-17
VF-211	NP 104	Nickel	CVA-31	CVW-21	AIM-9D	2126/10628	MiG-17
VF-211	NP 101	Nickel	CVA-31	CVW-21	AIM-9D	2050/10540	MiG-17
VF-211	NP	Nickel	CVA-31	CVW-21	AIM-9D*	2050/10540	MiG-17
VF-24	NP 4xx	Page Boy	CVA-31	CVW-21	AIM-9D	2050/10540	MiG-17
VF-24	NP 405	Page Boy	CVA-31	CVW-21	AIM-9D	2050/10540	MiG-17
VF-24	NP 442	Page Boy	CVA-31	CVW-21	AIM-9D	2103/10619	MiG-17
VF-24	NP 4xx	Page Boy	CVA-31	CVW-21	AIM-9	2103/10619	MiG-17 (P)
VF-24	NP 424	Page Boy	CVA-31	CVW-21	AIM-9/20mm	2103/10619	MiG-17
VF-211	NP 1xx	Nickel	CVA-31	CVW-21	20mm/Zuni	2118/10612	MiG-17D
VF-142	NK 202	Dakota	CVA-64	CVW-14	AIM-9	2038/10540	MiG-21
VF-142	NK 2xx	Dakota	CVA-64	CVW-14	AIM-9	2038-10540	MiG-21
VF-143	NK 1xx	Taproom	CVA-64	CVW-14	AIM-7	2045/10556	MiG-21
VF-142	NK 203	Dakota	CVA-64	CVW-14	AIM-7E	2110/10700	MiG-17

Squadron	Modex	Call Sign	CV	CVW	Weapon	Location	Kill
VF-162	AH 204	Superheat	CVA-34	CVW-16	AIM-9D	2045/10605	MiG-17
VF-96	NG 1xx	Showtime	CVN-65	CVW-9	AIM-7E		MiG-21 (P)*
VF-51	NL 116	Screaming Eagle	CVA-31	CVW-15	AIM-9	1855/10516	MiG-21
VF-191	NM 107	Feed Bag	CVA-14	CVW-19	AIM-9/ 20mm	1835/10530	MiG-17
VF-33	AE 212	Rootbeer	CVA-66	CVW-6	AIM-9	1845/10520	MiG-21
VF-53	NF 203	Firefighter	CVA-31	CVW-5	AIM-9	1856/10528	MiG-17
VF-51	NF 102	Screaming Eagle	CVA-31	CVW-5	AIM-9	1726/10612	MiG-21
VF-111	AK 103	Old Nick	CVS-11	CVW-10	AIM-9	1854/10521	MiG-21
VF-142	NK 201	Dakota	CVA-64	CVW-14	AIM-9		MiG-21
VF-96	NG 112	Showtime	CVA-64	CVW-9	AIM-9	1903/10517	MiG-21
VF-111	NL 201	Old Nick	CVA-43	CVW-15	AIM-9	1856/10503	MiG-17
VF-51	NL 100	Screaming Eagle	CVA-43	CVW-15	AIM-9	1946/10530	MiG-17
VF-114	NH 206	Linfield	CVA-63	CVW-11	AIM-9	2018/10538	MiG-21
VF-114	NH 201	Linfield	CVA-63	CVW-11	AIM-9	2018/10538	MiG-21
VF-96	NG 112	Showtime	CVA-64	CVW-9	AIM-9	2106/10521	MiG-17
VF-92	NG 211	Silver Kite	CVA-64	CVW-9	AIM-9	2127/10620	MiG-21F
VF-96	NG 106	Showtime	CVA-64	CVW-9	AIM-9	2057/10620	MiG-17
VF-96	NG 106	Showtime	CVA-64	CVW-9	AIM-9	2057/10620	MiG-17
VF-51	NL 111	Screaming Eagle	CVA-43	CVW-15	AIM-9	2053/10559	MiG-17
VF-96	NG 111	Showtime	CVA-64	CVW-9	AIM-9	2057/10620	MiG-17
VF-96	NG 100	Showtime	CVA-64	CVW-9	AIM-9	2055/10623	MiG-17
VF-96	NG 100	Showtime	CVA-64	CVW-9	AIM-9	2054/10620	MiG-17
VF-96	NG 100	Showtime	CVA-64	CVW-9	AIM-9	2053/10622	MiG-17
VF-161	NF 110	Rock River	CVA-41	CVW-5	AIM-9	2110/10630	MiG-19
VF-161	NF 105	Rock River	CVA-41	CVW-5	AIM-9G	2110/10630	MiG-19

Squadron	Modex	Call Sign	CV	CVW	Weapon	Location	Kill
VF-161	NF 100	Rock River	CVA-41	CVW-5	AIM 9	2125/10615	MiG-17
VF-161	NF 100	Rock River	CVA-41	CVW-5	AIM-9	2125/10615	MiG-17
VF-51	NL 114	Screaming Eagle	CVA-43	CVW-15	AIM-9	2032/10555	MiG-17
VF-51	NL 113	Screaming Eagle	CVA-43	CVW-15	AIM-9	2032/10555	MiG-17
VF-31	AC 101	Bandwagon	CVA-60	CVW-3	AIM-9	2125/10644	MiG-21
VF-103	AC 296	Clubleaf	CVA-60	CVW-3	AIM-7E	1930/10530	MiG-21J
VMFA-333	AJ 201	Shamrock	CVA-66	CVW-8	AIM-9G	2113/10549	MiG-21
VF-142	NK 214	Dakota	CVN-65	CVW-14	AIM-9	2057/10553	MiG-21
VF-161	NF 102	Rock River	CVA-41	CVW-5	AIM-9	2027/10713	MiG-17
13 TFS	OC	Gambit 03			AIM-4D		MiG-17
58 TFS	ZF	Dodge 01			AIM-7		MiG-21

J

Naval Aviators in Space

SOME MILESTONES

1. Alan B. Shepard, Jr., was the first American in space (suborbital flight), Freedom 7, 5 May 1961.
2. John H. Glenn, Jr., was the first American to orbit the earth, Friendship 7, 20 February 1962.
3. James A. Lovell, Jr., was one of three crewmen on the first flight to the moon.
4. Neil A. Armstrong was the first man to walk on the moon, Apollo 8, 21–27 December 1968.
5. The first U.S.-manned orbiting space station had an all-naval-aviation crew made up of Charles Conrad, Jr., Joseph P. Kerwin, and Paul J. Weitz, Skylab 2, 22 June 1973 (Kerwin was a navy flight surgeon).
6. Vance D. Brand participated in the first joint manned mission involving rendezvous, docking, crew transfer, and undocking of U.S. and Soviet spacecraft, Apollo-Soyuz Test Project, 15–24 July 1975.
7. The first flight into space by the Space Shuttle had an all-naval-

aviator crew made up of John W. Young and Robert L. Crippen, Space Shuttle Columbia, 12–14 April 1981.
8. Bruce McCandless made the first untethered space walk from Space Shuttle Challenger in February 1984 using a nitrogen-propelled manned maneuvering unit.

U.S. NAVY AND MARINE CORPS ASTRONAUTS*

The astronauts listed below were on active duty at the time of their selection for the space program or had a prior connection with navy or marine corps aviation. Most were naval aviators.

Astronauts	Affiliation
Armstrong, Neil A.	USN
Bagian, James P.	USN
Bean, Alan L.	USN
Bolden, Charles F.	USMC
Brand, Vance D.	USMC
Brandenstein, Daniel C.	USN
Buchli, James F.	USMC
Bull, John S.	USN
Cameron, Kenneth D.	USMC
Carpenter, M. Scott	USN
Carr, Gerald P.	USMC
Carter, Manley L., Jr.	USN
Cernan, Eugene A.	USN
Chaffee, Roger B.	USN
Coats, Michael L.	USN
Conrad, Charles, Jr.	USN
Creighton, John O.	USN
Crippen, Robert L.	USN
Culbertson, Frank L., Jr.	USN
Cunningham, Walter R.	USMC
Evans, Ronald E.	USN
Gardner, Dale A.	USN
Gibson, Robert L.	USN
Glenn, John H., Jr.	USMC
Gordon, Richard F., Jr.	USN
Griggs, S. David	USN
Haise, Fred W., Jr.	USMC
Hauck, Frederick H.	USN
Hilmers, David C.	USMC
Kerwin, Joseph P.	USN
Leestma, David C.	USN
Lind, Don L.	USN
Lounge, John M.	USN
Lousma, Jack R.	USMC

*List covers period from beginning of manned space program through May 1984.

Astronauts	*Affiliation*
Lovell, James A., Jr.	USN
Mattingly, Thomas K.	USN
McBride, Jon A.	USN
McCandless, Bruce, II	USN
McCulley, Michael J.	USN
Mitchell, Edgar D.	USN
O'Connor, Bryan D.	USMC
Overmyer, Robert F.	USMC
Richards, Richard N.	USN
Schirra, Walter M., Jr.	USN
See, Elliott M., Jr.	USN
Shepard, Alan B.	USN
Shepherd, William M.	USN
Smith, Michael J.	USN
Springer, Robert C.	USMC
Thagard, Norman E.	USMC
Truly, Richard H.	USN
Van Hoften, James D.	USN
Walker, David M.	USN
Weitz, Paul J.	USN
Wetherbee, James D.	USN
Williams, Clifton C.	USMC
Williams, Donald E.	USN
Young, John W.	USN

K

Naval Aviation Organizations

ASSOCIATIONS

Association of Naval Aviation, Inc. (ANA)

Dedicated to the support of naval aviation and a strong military posture. Membership is open to civilians and to military personnel of all services, all ranks and ratings, regular or reserve, active or retired. An annual convention alternates between East and West Coast locations. (5205 Leesburg Pike, Suite 200, Falls Church, Virginia 22401, [703] 998-7733)

Association of Naval Aviation

The Early and Pioneer Naval Aviators Association (Golden Eagles)

Dedicated to preserving the bonds of the past and fostering a pioneer spirit among present and future naval aviators. Membership is by invitation only and is made up of leading figures from naval aviation. The organization is strictly limited to 200 members. An annual convention is held each spring. (Captain E. Scott McCusky, USN, [Ret.], P.O. Box 7505, Clearwater, Florida 33575)

Golden Eagles

The Early and Pioneer Naval Aviators Association

Marine Corps Aviation Association (MCAA)

Dedicated to perpetuating the spirit of comradeship in marine aviation, fostering and encouraging professional excellence, and recognizing noteworthy achievements. Membership is open to marines or members of the armed forces, past and present, who have served with marine corps aviation units. (P.O. Box 296, Quantico, Virginia 22134, [703] 640-6161)

Marine Corps Aviation Association

Navy Helicopter Association (NHA)

Dedicated to the promotion and recognition of helicopters in naval aviation and to keeping members informed of new developments and accomplishments in the field. Membership is open to navy, coast guard, and marine corps aircrewmen and support personnel who are active, reserve, or retired. Annual

Navy Helicopter Association

convention. (P.O. Box 460, Coronado, California 92118, [714] 437-5708)

The Ancient Order of the Pterodactyl (USCG)

Dedicated to the promotion of social contact and camaraderie among coast guard aviators and supporters, maintenance of dialogue between present and past members of the coast guard, support of coast guard aviation and its goals, and recognition of the history of coast guard aviation. Membership is open to all who are serving or who have served honorably as pilots in coast guard aircraft, including those of other military services and foreign governments involved in exchange programs with the coast guard. Associate memberships are open to those who have served in other capacities aboard coast guard aircraft such as aircrewmen, flight surgeons, or others under official flight orders. Annual meeting. (P.O. Box 3133, Seal Beach, California 90740, [714] 963-8554)

The Ancient Order of the Pterodactyls

Silver Eagles Association, Inc.

Dedicated to fellowship among naval aviation pilots and to the perpetuation of the prominent role they played in the pioneering, development, and progress of naval aviation. Membership is open to all former and present enlisted pilots of the navy, marine corps, and coast guard. Annual convention. (2002 N. "D" Street, Pensacola, Florida 32501)

Silver Eagles Association

The Tailhook Association

Dedicated to the encouragement, support, and education of those interested in the aircraft carrier, her aircraft, pilots, and aircrewmen. Membership is open to anyone who has made an arrested carrier landing, as either a pilot or aircrew member. Associate membership is open to anyone who wants to support U.S. Navy carrier aviation. Annual reunion. (P.O. Box 40, Bonita, California 92002, [714] 479-8525)

The Tailhook Association

MUSEUMS

There are a number of museums throughout the United States that have naval aviation exhibits. The best known of these is the National Air and Space Museum, Smithsonian Institution, Washington, D.C. The Sea-Air Hall exhibit there has a simulated aircraft carrier complete with a hangar deck and World War II aircraft. The visitor may experience the sights and sounds of continuous carrier operations as seen from the island through an imaginative cinematic display and audio system.

There are four museums devoted exclusively to naval and marine corps aviation. The Naval Aviation Museum, at the air station in Pensacola, is the official museum of U.S. naval aviation. It features a large number of naval aircraft and aviation exhibits that trace the history of naval aviation from 1911 to the present. All significant examples of naval aircraft are represented, from a replica of the navy's first aeroplane to a modern space vehicle. An especially impressive restoration is the huge navy Curtiss NC-4, the first aircraft to negotiate the Atlantic by air in 1919. The museum is also the home of the Naval Aviation Hall of Honor, which honors the great names of naval aviation. The Naval Aviation Museum is open to the public from 9:00 AM to 5:00 PM daily.

The Marine Corps Aviation Museum in Quantico, Virginia, has a large number of historic marine corps aircraft dating from the earliest years. Housed in vintage hangars on the site of what was once Brown Field, the Marine Corps Aviation Museum is open to the public Tuesday through Friday, 10:00 AM to 5:00 PM, and weekends, 10:00 AM to 6:00 PM, April through November.

The USS *Yorktown* (CV 10) is the centerpiece of the Patriots Point Navy and Maritime Museum in Charleston, South Carolina. Acquired in 1975 and opened for visitors in 1976, the *Yorktown* has served as a memorial to carrier aviation. It has aircraft and other aviation exhibits and is the home of the Carrier Aviation Hall of Fame, which honors those who have made significant contributions to this form of naval warfare. The *Yorktown* is open to the public daily, 9:00 AM to 5:00 PM.

The USS *Intrepid Museum*, pier 86, is in New York, New York. The *Intrepid* (CV 11) is the second aircraft carrier to become a museum. It was acquired in 1980 and opened to the public on 3 August 1982. The ship contains aircraft static displays and exhibits that trace the evolution of the aircraft carrier to the modern carrier battle group. The museum is located on the west side of Manhattan and is open daily from 10:00 AM to 7:00 PM.

NAVY FLYING CLUBS

Navy flying clubs, located across the United States and abroad, give navy and marine corps personnel and their families an opportunity to develop flying skills and to become more closely involved with aviation. Membership is available to active-duty U.S. military personnel. Associate membership is open to dependents of active-duty personnel, retired U.S. military personnel and their dependents, full-time civilian employees of the Department of Defense, and members of the U.S. armed forces reserve. For information contact Recreational Services Department, Navy Military Personnel Command (N-116), Washington, D.C. 20370, (202) 694-3931.

NAVY FLYING CLUB LOCATIONS (as of 1 October 1984)

NAS Agana, Guam
NAS Atlanta, Georgia
NAS Atsugi, Japan
NAS Barbers Point, Hawaii

NWC China Lake, California
NAS Cubi Point, Philippines
NSWC Dahlgren, Virginia
NAS Dallas, Texas
NAF El Centro, California
NAS Glenview, Illinois
NSB New London, Connecticut
NS Guantanamo Bay, Cuba
NAS Jacksonville, Florida
Navy/Marine Reserve Center, Kansas City, Missouri
NS Keflavik, Iceland
NAS Key West, Florida
NAEC Lakehurst, New Jersey
NAS Lemoore, California
NAS Memphis, Tennessee
NAS Moffett Field, California
NPS Monterey, California
NSA Naples, Italy
USNA Annapolis, Maryland
NAS New Orleans, Louisiana
NETC Newport, Rhode Island
NAS Norfolk, Virginia
NAS North Island, California
NAC Orlando, Florida
NAS Patuxent River, Maryland
NAS Point Mugu, California
NS Roosevelt Roads, Puerto Rico
NS Rota, Spain
NAPC Trenton, New Jersey
NARC Twin Cities, Minnesota
NADC Warminster, Pennsylvania
NAS Whidbey Island, Washington

L

Naval Aviation Periodicals

There are many naval aviation-related periodicals. Air stations, aviation ships, and a myriad of other naval aviation activities publish newspapers and newsletters, as do reunion groups and a variety of naval aviation associations. Listed below are several periodicals generally available.

Approach—Magazine. Published monthly by the Naval Safety Center, NAS Norfolk, VA 23511. Deals primarily with flight safety. *Approach* is an internal publication of the U.S. Navy and is distributed without charge to squadrons, ships, and other naval aviation activities on the basis of 1 magazine for every 10 persons assigned. Also available from the superintendent of documents, U.S. Government Printing Office, Washington, D.C. 20402.

Flightlines—Newsletter. Published periodically by Coast Guard Headquarters, 2100 2nd Street S.W., Washington, D.C. 20593. It contains safety articles and other information of primary interest to the Coast Guard aviation community. *Flight Lines*, an internal publication of the U.S. Coast Guard, is distributed without charge to coast guard aviation activities.

Foundation—Magazine. Published semiannually by the Naval Aviation Museum Foundation Inc., NAS Pensacola, FL 32508. Contains articles on naval aviation history and information on the activities of the Naval Aviation Museum. *Foundation* is distributed to members of the Naval Aviation Museum Foundation as a benefit of membership and is also available through subscription.

Gold Book of Naval Aviation—Yearbook. Published annually for the Association of Naval Aviation, Inc., Suite 200, 5205 Leesburg Pike, Falls Church, VA 22041. Contains articles by ranking officers and other distinguished persons associated with naval aviation. Also contains current information and statistical data on a variety of naval aviation subjects. *The Gold Book* is available to members of the Association of Naval Aviation and the Naval Aviation Museum Foundation as a benefit of membership and is also for sale to the general public.

Mech—Magazine. Published bimonthly by the Naval Safety Center, NAS Norfolk, VA 23511. Oriented toward naval aviation maintenance personnel and contains articles and information pertaining to maintenance safety. *Mech* is an internal publication of the U.S. Navy and is distributed without charge to squadrons, ships, and other naval aviation activities on the basis of 1 magazine for every 10 persons assigned. Also available from the superintendent of documents, U.S. Government Printing Office, Washington, D.C. 20402.

Naval Aviation News—Magazine. Published bimonthly as a joint undertaking of the DCNO (air warfare) and the commander, Naval Air Systems Command. Serves as a communications link between these command structures on the one hand and the naval aviation community on the other. Contains articles and information of current and historical interest to naval aviation personnel at all levels. *Naval Aviation News* is an internal publication of the U.S. Navy and is distributed without charge to squadrons, ships, and other naval aviation activities on the basis of 1 magazine for every 7 persons assigned. Also available from the superintendent of documents, U.S. Government Printing Office, Washington, D.C. 20402.

Proceedings of the U.S. Naval Institute—Magazine. Published monthly by the U.S. Naval Institute, Annapolis, MD 21402. Contains articles and information on subjects of naval interest, including many on various aspects of naval aviation. *Proceedings* is distributed to members of the Naval Institute as a benefit of membership and is also available by subscription.

Rotor Review—Newsletter. Published periodically by the Navy Helicopter Association, P.O. Box 460, Coronado, CA 92118. Contains information of interest to the navy helicopter community and the activities of the association. *Rotor Review* is distributed to the members of the Navy Helicopter Association as a benefit of membership.

The Hook—Magazine. Published quarterly by The Tailhook Association, P.O. Box 40, Bonita, CA 92002. Contains articles and information on carrier operations past and present with a number of pages devoted to current activities of today's carrier air wings. *The Hook* is distributed to members as a benefit of membership and is also available through subscription.

Wings of Gold—Magazine. Published quarterly by the Association of Naval Aviation Inc., Suite 200, 5205 Leesburg Pike, Falls Church, VA 22041. Deals primarily with current aspects of naval aviation. Provides a forum for an exchange of views from various perspectives. Also contains some articles on naval aviation history. *Wings of Gold* is distributed to members of the Association of Naval Aviation as a benefit of membership and is also available through subscription.

Yellow Sheet—Newsletter. Published quarterly by the Marine Corps Aviation Association, P.O. Box 296, Quantico, VA 22134. Contains articles and information on Marine Corps aviation and association activities. *Yellow Sheet* is distributed to members of the Marine Corps Aviation Association as a benefit of membership.

M

Acronyms and Abbreviations Commonly Used in Naval Aviation

A/A   Angle of attack
AAM   Air-to-air missile
A/C   Aircraft
ACDUTRA   Active duty for training
AGL   Above-ground level
ACIP   Aviation career incentive pay
ACLS   Automatic carrier landing system
ACM   Air combat maneuver
ADIZ   Air defense identification zone
ADPO   Advanced development program office
AEDO   Aeronautical engineering duty officer
AEW   Airborne early warning
AEWWING   Airborne early warning wing
AGM   Aircraft ground mishap
AIMD   Aircraft intermediate-maintenance department
AIMSO   Aircraft Intermediate-Support Office
AIO   Air intelligence officer
ALF   Auxiliary landing field
ALT   Altitude

AMB   Aircraft mishap board
AMDO   Aeronautical maintenance duty officer
AMO   Aircraft maintenance officer
AMSL   Above mean sea level
ANA   Association of Naval Aviation
AOCC   Air operations control center (amphibious operations)
AOC   Aviation officer candidate
AOCP   Aviation officer continuation pay
AOM   All officers meeting
APM   All pilots meeting
APC   Approach control
APU   Auxiliary power unit
ARREC   Accident report recommendations
ASCAC   Antisubmarine classification and analysis center
ASM   Air-to-surface missile
ASO   Aviation safety officer, Aviation Supply Office, aviation supply officer
ASR   Air surveillance radar
ASW   Antisubmarine warfare
ASWEX   Antisubmarine warfare exercise
ATC   Air traffic control
ATDS   Airborne tactical data system
AUW   Advanced underseas weapons
AVCAL   Aviation consolidated-allowance list
AVT   Aviation training ship
BALL   Optical landing aid
BINGO   Proceed and land at field specified
BIS   Board of Inspection and Survey
B/N   Bombardier/navigator
BOLTER   Takeoff following unsuccessful arrestment
BRC   Base recovery course
BRG   Bearing
BUNO   Bureau number
CAG   Commander air wing
CANTRAC   Catalog of navy training courses
CARGRU   Carrier group
CATCC   Carrier air-traffic-control center

CAS   Calibrated air speed
CAVU   Ceiling and visibility unlimited
CCA   Carrier controlled approach
CEP   Circular error probable
CGAS   Coast guard air station
CGATC   Coast guard aviation training center
CH   Channel
CHOP   Change of operational control
CIC   Combat information center
CINC   Commander in chief
CLARA   Ball not in sight
CNATECHTRA   Chief of Naval Air Technical Training
CNATRA   Chief of Naval Air Training
CNAVRES   Chief of Naval Reserve
CNET   Chief of naval education and training
CO   Commanding officer
COD   Carrier onboard delivery
COMASWWINGPAC   Commander antisubmarine warfare wing, Pacific
COMCARGRU   Commander carrier group
COMCVWR   Commander carrier reserve wing
COMFAIR   Commander fleet air (followed by location)
COMFITAEWWINGPAC   Commander Fighter Airborne Early Warning Wing, Pacific
COMHELWINGRES   Commander Helicopter Reserve Wing
COMLATWINGPAC   Commander Light-Attack Wing, Pacific
COMMATVAQWINGPAC   Commander Medium-Attack Electronic Warfare Wing, Pacific
COMNAVAIRLANT   Commander Naval Air Forces, Atlantic
COMNAVAIRPAC   Commander Naval Air Forces, Pacific
COMNAVAIRRESFOR   Commander Naval Air Reserve Force
COMNAVAIRSYSCOM   Commander Naval Air Systems Command
COMOPTEVFOR   Commander Operational Test and Evaluation Force
COMPATWING   Commander patrol wing (followed by a numeral)

COMPATWINGPAC   Commander Patrol Wing, Pacific
COMRESPATWING   Commander reserve patrol wing
COMRESTACSUPPWING   Commander Reserve Tactical Support Wing
COMSEABASEDASWWINGSLANT   Commander Sea-based Antisubmarine Warfare Wings, Atlantic
COMTACSUPWING   Commander tactical support wing
COMTACWINGSLANT   Commander Tactical Wings, Atlantic
CONUS   Continental United States
CTF   Commander task force
CV   Multipurpose aircraft carrier
CVN   Multipurpose aircraft carrier (nuclear)
CVW   Carrier air wing
CVWR   Carrier air wing reserve
DELTA   Hold and conserve fuel signal
DET   Detachment
DCNO   Deputy Chief of Naval Operations
DIFDEN   Duty in a flying status not involving flying
DIFOPS   Duty in a flying status involving operational or training flights
DME   Distance-measuring equipment
DOD   Department of Defense
DON   Department of the Navy
EAT   Expected approach time
ECM   Electronic countermeasures
EMCON   Emission control
ETA   Estimated time of arrival
ETD   Estimated time of departure
ETE   Estimated time en route
EW   Electronic warfare
FAA   Federal Aviation Agency
FAF   Final-approach fix
FAM   Familiarization
FCLP   Field carrier-landing practice
FFEB   Field flight evaluation board
FITWEPSCOL   Fighter Weapons School

FITWING   Fighter wing
FL   Flight level
FLDO   Flying limited duty officer
FLECOMPRON   Fleet composite squadron
FLIP   Flight-information publication
FLIR   Forward-looking infrared radar
FLOLS   Fresnel lens optical-landing system
FM   Flight mishap
FMF   Fleet Marine Force
FNAEB   Field naval aviator evaluation board
FOD   Foreign object damage
FPM   Feet per minute
FRAMP   Fleet readiness/replacement aviation maintenance
     personnel
FRM   Flight-related mishap
FRS   Fleet readiness squadron
FSR   Flight surgeon's report
GCA   Ground control approach
GS   Ground speed
GSE   Ground support equipment
HAC   Helicopter aircraft commander
HAL   Helicopter attack squadron light
HC   Helicopter combat-support squadron
HDC   Helicopters direction center (amphibious operations)
HELSEACONTROLWING   Helicopter sea-control wing
HERO   Hazard of electronic radiation to ordnance
HF   High frequency
H&HS   Headquarters and headquarters squadron (marine)
HM   Helicopter minecountermeasures squadron
HMA   Marine helicopter attack squadron
HMH   Marine heavy-helicopter squadron
HML   Marine light-helicopter squadron
HMM   Marine medium-helicopter squadron
HMT   Marine helicopter training squadron
HMX   Marine helicopter squadron (presidential transport)
HQMC   Headquarters Marine Corps
HR   Hazard report

HS   Helicopter antisubmarine squadron
HSWING   Helicopter antisubmarine wing
HSL   Helicopter antisubmarine squadron light
HT   Helicopter training squadron
HUD   Heads-up display
IAF   Initial-approach fix
IAS   Indicated air speed
ICAO   International Civil Aviation Organization
ICBM   Intercontinental ballistic missile
ICLS   Instrument carrier-landing system
IFF   Identification friend or foe
ILS   Instrument landing system
IMA   Intermediate maintenance activities
IMC   Instrument meteorological conditions
JATO   Jet-assisted takeoff
JCN   Job control number
KIAS   Knots-indicated air speed
KILO   Pilot report indicating aircraft mission readiness
LAMPS   Light airborne multipurpose system
LASER   Light amplification by simulated emission of radiation
LATWING   Light-attack wing
LDG   Landing
LF   Low frequency
LHA   Helicopter-assault landing ship
LLD   Light landing device
LORAN   Long-range aid to navigation
LOS   Line of sight
LOX   Liquid oxygen
LPH   Amphibious-assault ship
LSE   Landing signal enlisted
LSO   Landing signal officer
MAB   Marine air-base squadron
MACG   Marine air-control group
MACS   Marine air-control squadron
MAD   Magnetic anomaly detection (equipment)
MAD   Marine aviation detachment
MAG   Marine aircraft group

MARTD   Marine air reserve training detachment
MASS   Marine air support squadron
MATCS   Marine air-traffic-control squadron
MATCU   Marine air-traffic-control unit
MATSG   Marine air-training support group
MATWING   Medium-attack wing
MAW   Marine aircraft wing
MAWTS-1   Marine aviation weapons and tactics squadron 1
MAYDAY   International aircraft distress signal
MC   Misson control
MCAF   Marine corps air facility
MCAS   Marine corps air station
MCM   Mine countermeasures
MEDEVAC   Medical evacuation
MIM   Maintenance instruction manual
MIR   Mishap investigation report
MOVLAS   Manually operated visual-landing system
MR   Mishap report
MSL   Mean sea level
MWCS   Marine wing communication squadron
MWHS   Marine wing headquarters squadron
MWSG   Marine wing support group
NA   Naval aviator
NAAF   Naval auxiliary airfield
NAC   Naval aircrewman, Naval Avionics Center
NADC   Naval Air Development Center
NAEC   Naval Aviation Engineering Center
NAESU   Naval aviation engineering service unit
NAF   Naval air facility
NALCOMIS   Naval aviation logistics-command management-
    information system
NAMI   Naval Aerospace Medical Institute
NAMP   Naval aviation maintenance program
NAMRL   Naval Aerospace Medical Research Laboratory
NAMSO   Navy Maintenance Support Office
NAMTRAGRU   Naval air maintenance training group
NAPC   Naval Air Propulsion Center

NAR   Naval air reserve
NAS   Naval air station
NASA   National Aeronautics and Space Administration
NATC   Naval Air Test Center
NATOPS   Naval air training and operating-procedures stand-
    ardization
NATSF   Naval Air Technical Services Facility
NATTC   Naval Air Technical Training Center
NAVAIRDEVCEN   Naval Air Development Center
NAVAIRENGCEN   Naval Air Engineering Center
NAVAIRREWORKFAC   Naval air rework facility
NAVAIRSYSCOM   Naval Air Systems Command
NAVAVNLOGCEN   Naval Aviation Logistics Center
NAVAVNWEPSFAC   Naval aviation weapons facility
NAVFAC   Naval facility
NAVFITWEPSCOL   Navy Fighter Weapons School
NAVPRO   Naval plant representative office
NAVSAFECEN   Naval Safety Center
NAVSTA   Naval station
NAVSTKWARCEN   Naval Strike Warfare Center
NAVWESA   Naval Weapons Engineering Support Activity
NAVWPNEVALFAC   Naval Weapons Evaluation Facility
NEC   Navy enlisted classification
NEPRF   Naval Environmental-Prediction Research Facility
NFO   Naval flight officer
NMAC   Near midair collision
NMCS   Not-mission-capable supply
NOAA   National Oceanic and Atmospheric Administration
NOBC   Naval officer billet code
NOTAM   Notice to airmen
NOTS   Naval ordnance test station
NTDS   Navy tactical data system
NWP   Naval warfare publication
NWS   Naval weapons station
OFT   Operational flight trainer
OINC   Officer in charge
OJN   Overwater jet navigation

OLF   Outlying field
O&MN   Operations and maintenance navy
OOD   Officer of the deck
OP-05   Deputy chief of naval operations (air warfare)
OPNAV   Office of the Chief of Naval Operations
OPTAR   Operational target (funding)
OPTEVFOR   Operational Test and Evaluation Force
ORE   Operational-readiness evaluation
OSAP   Ocean-surveillance air patrol
OTC   Officer in tactical command
PAR   Precision-approach radar
PATRON   Patrol squadron
PATWING   Patrol wing
PAX   Passenger
PIM   Position and intended movement
PM   Periodic maintenance or preventive maintenance
PMA   Program manager air
PMC   Passenger/mail/cargo (logistics flight)
PMCF   Post-maintenance-check flight
PMCS   Partial-mission-capable supply
PMRF   Pacific missile range facility
PMTC   Pacific Missile Test Center
POSIT   Position
PPC   Patrol-plane commander
QA   Quality assurance
QEC   Quick engine change
RAC   Risk assessment code
RADHAZ   Radiation hazards
RATCC   Radar air-traffic-control center
RAST   Recovery assist secure and traverse system (small deck helicopter recovery)
RDT&E   Research, development, test, and evaluation
RECCE/RECON   Reconnaissance
RF   Radio frequency
RFF   Ready for ferry
RFI   Ready for issue
RIO   Radar intercept officer

ROE   Rules of engagement
RON   Remain overnight
RPM   Revolutions per minute
RVL   Rolling vertical landing (V/STOL)
RVTO   Rolling vertical takeoff (V/STOL)
RWY   Runway
SAM   Surface-to-air missile
SAR   Search and rescue, or selected air reservist
SAU   Squadron augmentation unit
SDO   Squadron duty officer
SECNAV   Secretary of the navy
SF   Standard form
SIGINT   Signal intelligence
SLBM   Sea-launched ballistic missile
SLCM   Sea-launched cruise missile
SLEP   Service life extension program
SNDL   Standard navy distribution list
SONOBUOY   Underwater listening buoy for aircraft use
SUPO   Supply officer
SYSCOM   Systems command
TACAIR   Tactical air
TACAMO   Take action and move out
TACAN   Tactical air navigation (system)
TACCO   Tactical coordinator
TACGRU   Tactical air control group
TACRON   Tactical air control squadron
TACSUPWING   Tactical support wing
TACTS   Tactical aircrew combat training system
TAD/TEMADD   Temporary additional duty
TAR   Training and administration of reserves
TARPS   Tactical air reconnaissance pod system
TAS   True air speed
TAT   Turnaround time
TD   Technical directive
TDY/TEMDU   Temporary duty
TIT   Turbine inlet temperature
TN   Tactical navigation

TO   Takeoff
TRACOM   Training Command
TRAP   Successful arrestment
TRARON   Training squadron
TRAWING   Training wing
UHF   Ultra-high frequency
UIC   Unit identification code
UNCLAS   Unclassified
UNK   Unknown
VA   Attack squadron
VAK   Tactical aerial-refueling squadron
VAQ   Tactical electronic-warfare squadron
VAW   Carrier airborne-early-warning squadron
VC   Fleet composite squadron
VERTREP   Vertical replenishment
VF   Fighter squadron
VFA   Strike-fighter squadron
VFR   Visual flight rules
VHF   Very-high frequency
VMA   Marine attack squadron
VMAT(AW)   Marine all-weather attack squadron
VMAQ   Marine tactical/electronic-warfare squadron
VMAT   Marine attack training squadron
VMAT(AW)   Marine all-weather attack training squadron
VMC   Visual meteorological conditions
VMFA   Marine fighter/attack squadron
VMFAT   Marine fighter/attack training squadron
VMFP   Marine tactical reconnaissance squadron
VMGR   Marine aerial refueler/transport squadron
VMO   Marine observation squadron
VOR   Visual omni range
VP   Patrol squadron
VQ   Fleet air reconnaissance squadron
VR   Aircraft logistics support squadron
VRC   Aircraft logistics support squadron (COD)
VRF   Aircraft ferry squadron
VS   Carrier antisubmarine-warfare squadron

V/STOL   Vertical/short takeoff and landing
VT   Training squadron
VTOL/VERTOL   Vertical takeoff and landing
VX   Air test and evaluation squadron
VXE   Antarctic development squadron
VXN   Oceanographic development squadron
W&B   Weight and balance
WC   Work center
WEC   Wing engineer squadron (marine)
WOD   Wind over the deck
WTS   Wing transportation squadron (marine)
WX   Weather
X   Experimental
XMIT   Transmit
XO   Executive officer
ZIP-LIP   Communications minimized
ZT   Time zone

Index

About the Authors and Revisers

Dr. William J. Armstrong—historian and archivist for the Naval Air Systems Command and author of numerous articles on naval aviation subjects.

Commander Jess C. Barrow, USCGR (Ret.)—naval aviator and historian, and author of several articles on coast guard aviation and a book on marine corps aviation.

Captain Paolo Coletta, USNR (Ret.)—historian, former professor of history at the U.S. Naval Academy, authority on naval air stations and facilities, and author of numerous papers, articles, and books on naval subjects.

Captain J. J. Coonan, USN—naval aviator, former squadron and carrier air wing commander, former aviation officer community manager for the Naval Military Personnel Command, and currently commander of Light Attack Wing 1.

Major John M. Elliott, USMC (Ret.)—aviation historian, authority on aircraft restoration, magazine writer, and author of a comprehensive work on aircraft markings and paint schemes.

Chief Warrant Officer J. Emmert, USN—experienced aviation maintenance officer and currently maintenance/material control officer at the Naval Air Facility in Washington, D.C.

Journalist Senior Chief Kirby J. Harrison, USN—photojournalist, magazine and book editor, and author of numerous articles on naval aviation subjects.

Captain Richard C. Knott, USN—naval aviator, aviation historian, magazine and book editor, and author of two books on naval aviation subjects.

Commander Peter B. Mersky, USNR—aviation historian, active naval air reservist, magazine editor and illustrator, book review editor, and author of two books and numerous articles on naval aviation subjects.

Captain John B. Noll, MC, USN—flight surgeon, former assistant director of training at the Naval Aerospace Medical Institute, and currently assigned to the Naval Medical Command, Washington, D.C.

Captain Albert L. Raithel, USN (Ret.)—naval aviator, aviation writer and historian, and an authority on naval aviation in World War I.

Captain Rosario Rausa, USNR—naval aviator, assistant safety coordinator in the office of the CNO, former magazine editor, and author of numerous articles and several books on naval aviation subjects.

Lieutenant Peter J. Reynierse, USNR—public affairs officer and writer and researcher on the subject of naval aviation supply.

Commander Van N. Stewart, USN—naval aviator, former flight instructor in pilot and naval flight officer programs, former CNO program coordinator for the T-45 training system, and currently deputy program manager (readiness) for the T-45 training system, Naval Air Systems Command.

Commodore Jeremy D. Taylor, USN—naval aviator, former squadron and carrier air wing commander, and former commanding officer of the USS *Coral Sea*.

Mr. Robert H. Thompson—civilian staff officer in the Office of the Deputy CNO (air warfare) and an authority on naval organizational relationships.